Modeling Real Life

Grade 4

Volume 1

Ron Larson
Laurie Boswell

Erie, Pennsylvania
BigIdeasLearning.com

Big Ideas Learning, LLC
1762 Norcross Road
Erie, PA 16510-3838
USA

For product information and customer support, contact Big Ideas Learning
at 1-877-552-7766 or visit us at BigIdeasLearning.com.

Cover Image
Valdis Torms, Brazhnykov Andriy/Shutterstock.com

Printed in the U.S.A.

ISBN 13: 978-1-63598-889-5

5 6 7 8 9 10—22 21

About the Authors

Ron Larson

Ron Larson, Ph.D., is well known as the lead author of a comprehensive program for mathematics that spans school mathematics and college courses. He holds the distinction of Professor Emeritus from Penn State Erie, The Behrend College, where he taught for nearly 40 years. He received his Ph.D. in mathematics from the University of Colorado. Dr. Larson's numerous professional activities keep him actively involved in the mathematics education community and allow him to fully understand the needs of students, teachers, supervisors, and administrators.

Ron Larson

Laurie Boswell

Laurie Boswell, Ed.D., is the former Head of School at Riverside School in Lyndonville, Vermont. In addition to textbook authoring, she provides mathematics consulting and embedded coaching sessions. Dr. Boswell received her Ed.D. from the University of Vermont in 2010. She is a recipient of the Presidential Award for Excellence in Mathematics Teaching and is a Tandy Technology Scholar. Laurie has taught math to students at all levels, elementary through college. In addition, Laurie has served on the NCTM Board of Directors and as a Regional Director for NCSM. Along with Ron, Laurie has co-authored numerous math programs and has become a popular national speaker.

Laurie Boswell

Dr. Ron Larson and Dr. Laurie Boswell began writing together in 1992. Since that time, they have authored over four dozen textbooks. This successful collaboration allows for one voice from Kindergarten through Algebra 2.

Contributors, Reviewers, and Research

Big Ideas Learning would like to express our gratitude to the mathematics education and instruction experts who served as our advisory panel, contributing specialists, and reviewers during the writing of *Big Ideas Math: Modeling Real Life*. Their input was an invaluable asset during the development of this program.

Contributing Specialists and Reviewers

- **Sophie Murphy**, Ph.D. Candidate, Melbourne School of Education, Melbourne, Australia
  Learning Targets and Success Criteria Specialist and Visible Learning Reviewer
- **Linda Hall**, Mathematics Educational Consultant, Edmond, OK
  Advisory Panel
- **Michael McDowell**, Ed.D., Superintendent, Ross, CA
  Project-Based Learning Specialist
- **Kelly Byrne**, Math Supervisor and Coordinator of Data Analysis, Downingtown, PA
  Advisory Panel
- **Jean Carwin**, Math Specialist/TOSA, Snohomish, WA
  Advisory Panel
- **Nancy Siddens**, Independent Language Teaching Consultant, Las Cruces, NM
  English Language Learner Specialist
- **Kristen Karbon**, Curriculum and Assessment Coordinator, Troy, MI
  Advisory Panel
- **Kery Obradovich**, K–8 Math/Science Coordinator, Northbrook, IL
  Advisory Panel
- **Jennifer Rollins**, Math Curriculum Content Specialist, Golden, CO
  Advisory Panel
- **Becky Walker**, Ph.D., School Improvement Services Director, Green Bay, WI
  Advisory Panel and Content Reviewer
- **Deborah Donovan**, Mathematics Consultant, Lexington, SC
  Content Reviewer
- **Tom Muchlinski**, Ph.D., Mathematics Consultant, Plymouth, MN
  Content Reviewer and Teaching Edition Contributor
- **Mary Goetz**, Elementary School Teacher, Troy, MI
  Content Reviewer
- **Nanci N. Smith**, Ph.D., International Curriculum and Instruction Consultant, Peoria, AZ
  Teaching Edition Contributor
- **Robyn Seifert-Decker**, Mathematics Consultant, Grand Haven, MI
  Teaching Edition Contributor
- **Bonnie Spence**, Mathematics Education Specialist, Missoula, MT
  Teaching Edition Contributor
- **Suzy Gagnon**, Adjunct Instructor, University of New Hampshire, Portsmouth, NH
  Teaching Edition Contributor
- **Art Johnson**, Ed.D., Professor of Mathematics Education, Warwick, RI
  Teaching Edition Contributor
- **Anthony Smith**, Ph.D., Associate Professor, Associate Dean, University of Washington Bothell, Seattle, WA
  Reading and Writing Reviewer
- **Brianna Raygor**, Music Teacher, Fridley, MN
  Music Reviewer
- **Nicole Dimich Vagle**, Educator, Author, and Consultant, Hopkins, MN
  Assessment Reviewer
- **Janet Graham**, District Math Specialist, Manassas, VA
  Response to Intervention and Differentiated Instruction Reviewer
- **Sharon Huber**, Director of Elementary Mathematics, Chesapeake, VA
  Universal Design for Learning Reviewer

Student Reviewers

- T.J. Morin
- Alayna Morin
- Ethan Bauer
- Emery Bauer
- Emma Gaeta
- Ryan Gaeta
- Benjamin SanFrotello
- Bailey SanFrotello
- Samantha Grygier
- Robert Grygier IV
- Jacob Grygier
- Jessica Urso
- Ike Patton
- Jake Lobaugh
- Adam Fried
- Caroline Naser
- Charlotte Naser

Research

Ron Larson and Laurie Boswell used the latest in educational research, along with the body of knowledge collected from expert mathematics instructors, to develop the *Modeling Real Life* series. The pedagogical approach used in this program follows the best practices outlined in the most prominent and widely accepted educational research, including:

- *Visible Learning*
  John Hattie © 2009
- *Visible Learning for Teachers*
  John Hattie © 2012
- *Visible Learning for Mathematics*
  John Hattie © 2017
- *Principles to Actions: Ensuring Mathematical Success for All*
  NCTM © 2014
- *Adding It Up: Helping Children Learn Mathematics*
  National Research Council © 2001
- *Mathematical Mindsets: Unleashing Students' Potential through Creative Math, Inspiring Messages and Innovative Teaching*
  Jo Boaler © 2015
- *What Works in Schools: Translating Research into Action*
  Robert Marzano © 2003
- *Classroom Instruction That Works: Research-Based Strategies for Increasing Student Achievement*
  Marzano, Pickering, and Pollock © 2001
- *Principles and Standards for School Mathematics*
  NCTM © 2000
- *Rigorous PBL by Design: Three Shifts for Developing Confident and Competent Learners*
  Michael McDowell © 2017
- *Universal Design for Learning Guidelines*
  CAST © 2011
- Rigor/Relevance Framework®
  International Center for Leadership in Education
- *Understanding by Design*
  Grant Wiggins and Jay McTighe © 2005
- Achieve, ACT, and The College Board
- *Elementary and Middle School Mathematics: Teaching Developmentally*
  John A. Van de Walle and Karen S. Karp © 2015
- *Evaluating the Quality of Learning: The SOLO Taxonomy*
  John B. Biggs & Kevin F. Collis © 1982
- *Unlocking Formative Assessment: Practical Strategies for Enhancing Students' Learning in the Primary and Intermediate Classroom*
  Shirley Clarke, Helen Timperley, and John Hattie © 2004
- *Formative Assessment in the Secondary Classroom*
  Shirley Clarke © 2005
- *Improving Student Achievement: A Practical Guide to Assessment for Learning*
  Toni Glasson © 2009

Mathematical Processes and Proficiencies

Big Ideas Math: Modeling Real Life reinforces the Process Standards from NCTM and the Five Strands of Mathematical Proficiency endorsed by the National Research Council. With *Big Ideas Math*, students get the practice they need to become well-rounded, mathematically proficient learners.

Problem Solving/Strategic Competence

- *Think & Grow: Modeling Real Life* examples use problem-solving strategies, such as drawing a picture, circling knowns, and underlining unknowns. They also use a formal problem-solving plan: understand the problem, make a plan, and solve and check.
- Real-life problems are provided to help students learn to apply the mathematics that they are learning to everyday life.
- Real-life problems help students use the structure of mathematics to break down and solve more difficult problems.

Reasoning and Proof/Adaptive Reasoning

- *Explore & Grows* allow students to investigate math and make conjectures.
- Questions ask students to explain and justify their reasoning.

Communication

- Cooperative learning opportunities support precise communication.
- Exercises, such as *You Be The Teacher* and *Which One Doesn't Belong?,* provide students the opportunity to critique the reasoning of others.
- *Apply and Grow: Practice* exercises allow students to demonstrate their understanding of the lesson up to that point.
- *ELL Support* notes provide insights into how to support English learners.

Connections

- Prior knowledge is continually brought back and tied in with current learning.
- Performance Tasks tie the topics of a chapter together into one extended task.
- Real-life problems incorporate other disciplines to help students see that math is used across content areas.

Representations/Productive Disposition

- Real-life problems are translated into pictures, diagrams, tables, equations, and graphs to help students analyze relations and to draw conclusions.
- Visual problem-solving models help students create a coherent representation of the problem.
- Multiple representations are presented to help students move from concrete to representative and into abstract thinking.
- *Learning Targets* and *Success Criteria* at the start of each chapter and lesson help students understand what they are going to learn.
- Real-life problems incorporate other disciplines to help students see that math is used across content areas.

Conceptual Understanding

- *Explore & Grows* allow students to investigate math to understand the reasoning behind the rules.

Procedural Fluency

- Skill exercises are provided to continually practice fundamental skills.
- Prior knowledge is continually brought back and tied in with current learning.

Meeting Proficiency and Major Topics

Meeting Proficiency

As standards shift to prepare students for college and careers, the importance of focus, coherence, and rigor continues to grow.

FOCUS *Big Ideas Math: Modeling Real Life* emphasizes a narrower and deeper curriculum, ensuring students spend their time on the major topics of each grade.

COHERENCE The program was developed around coherent progressions from Kindergarten through eighth grade, guaranteeing students develop and progress their foundational skills through the grades while maintaining a strong focus on the major topics.

RIGOR *Big Ideas Math: Modeling Real Life* uses a balance of procedural fluency, conceptual understanding, and real-life applications. Students develop conceptual understanding in every *Explore and Grow*, continue that development through the lesson while gaining procedural fluency during the *Think and Grow*, and then tie it all together with *Think and Grow: Modeling Real Life*. Every set of practice problems reflects this balance, giving students the rigorous practice they need to be college- and career-ready.

Major Topics in Grade 4

Operations and Algebraic Thinking

- Use the four operations with whole numbers to solve problems.

Number and Operations in Base Ten

- Generalize place value understanding for multi-digit whole numbers.
- Use place value understanding and properties of operations to perform multi-digit arithmetic.

Number and Operations—Fractions

- Extend understanding of fraction equivalence and ordering.
- Build fractions from unit fractions by applying and extending previous understandings of operations on whole numbers.
- Understand decimal notation for fractions, and compare decimal fractions.

Use the color-coded Table of Contents to determine where the major topics, supporting topics, and additional topics occur throughout the curriculum.

- Major Topic
- Supporting Topic
- Additional Topic

Place Value Concepts

Add and Subtract Multi-Digit Numbers

■ Major Topic
■ Supporting Topic
■ Additional Topic

Multiply by One-Digit Numbers

Multiplication Quest

Directions:

1. Players take turns rolling a die. Players solve problems on their boards to race the knights to their castles.
2. On your turn, solve the next multiplication problem in the row of your roll.
3. The first player to get a knight to a castle wins!

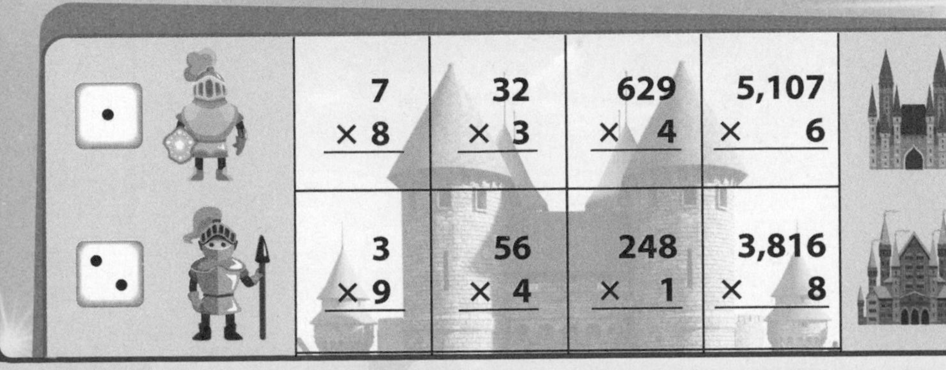

4 Multiply by Two-Digit Numbers

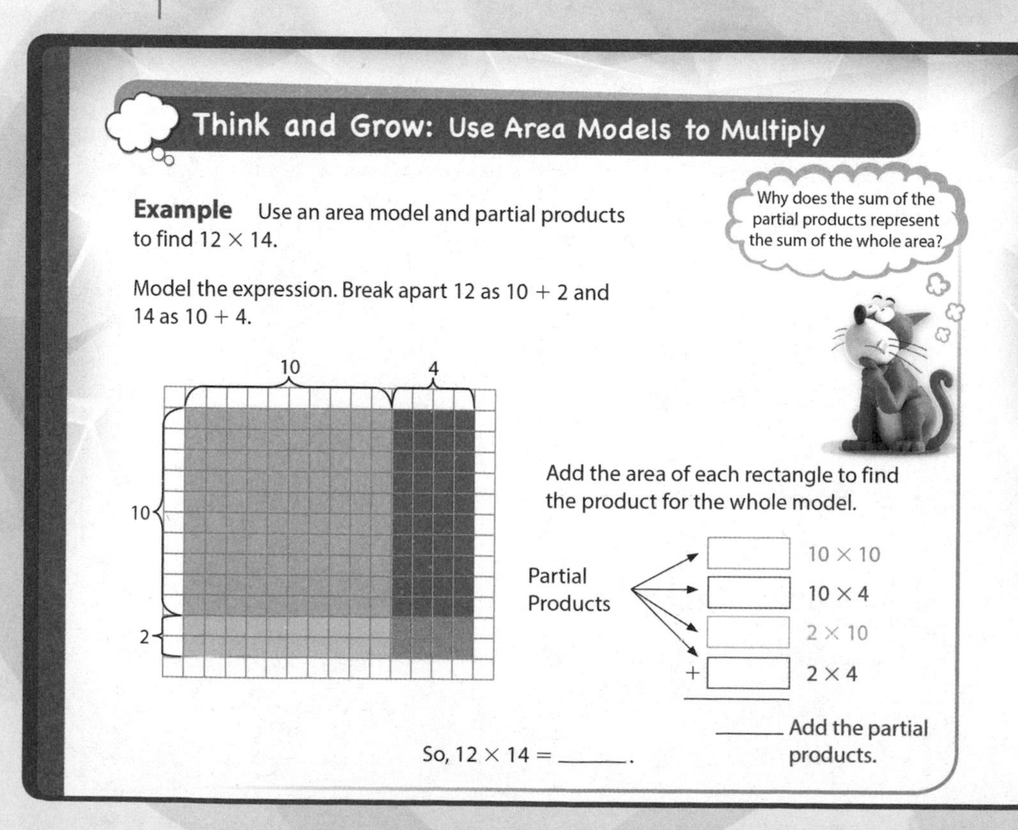

Divide Multi-Digit Numbers by One-Digit Numbers

Factors, Multiples, and Patterns

- Major Topic
- Supporting Topic
- Additional Topic

Understand Fraction Equivalence and Comparison

Add and Subtract Fractions

- Major Topic
- Supporting Topic
- Additional Topic

Multiply Whole Numbers and Fractions

10 Relate Fractions and Decimals

Let's learn how to relate fractions and decimals!

Understand Measurement Equivalence

12

Use Perimeter and Area Formulas

■ Major Topic
■ Supporting Topic
■ Additional Topic

Identify and Draw Lines and Angles

14 Identify Symmetry and Two-Dimensional Shapes

1 Place Value Concepts

- What kinds of numbers would you find on a map?
- Why is place value important when you read a map?

Chapter Learning Target:
Understand place value.

Chapter Success Criteria:
- I can define the value of a number.
- I can explain how to use symbols to compare two numbers.
- I can compare the value of two identical digits in a number.
- I can read and write multi-digit numbers in multiple forms.

Name ______________________

1 Vocabulary

Review Words
expanded form
standard form
word form

Organize It

Use the review words to complete the graphic organizer.

Ways to Write Numbers

[ ] three hundred fifty-seven

[ ] $300 + 50 + 7$

[ ] 357

Define It

Use your vocabulary cards to complete each definition.

1. period: Each group of ________ digits separated by ________ in a multi-digit ________
2. place value chart: A ________ that shows the ________ of each digit in a ________
3. thousands period: The period ________ the ones period in a number

Chapter 1 Vocabulary Cards

ones period

period

place value chart

thousands period

Each group of three digits separated by commas in a multi-digit number

period period

Thousands Period			Ones Period		
Hundreds	Tens	Ones	Hundreds	Tens	Ones
1	0	0,	0	0	0

The first period in a number

Thousands Period			Ones Period		
Hundreds	Tens	Ones	Hundreds	Tens	Ones
8	1	5,	7	9	6

The period after the ones period in a number

Thousands Period			Ones Period		
Hundreds	Tens	Ones	Hundreds	Tens	Ones
8	1	5,	7	9	6

A chart that shows the value of each digit in a number

Thousands Period			Ones Period		
Hundreds	Tens	Ones	Hundreds	Tens	Ones
2	8	5,	7	4	3

Name ____________________

Understand Place Value 1.1

Learning Target: Identify the values of digits in multi-digit numbers.

Success Criteria:

- I can identify the first six place value names.
- I can identify the value of each digit in a number.
- I can compare the values of two of the same digits in a number.

Model the number. Draw to show your model. Then write the value of each digit.

1,275

1000 + 200 + 70 + 5

3,333

_____ _____ _____ _____

Repeated Reasoning Compare the value of the tens digit to the value of the ones digit. Then do the same with the hundreds and tens digits, and the thousands and hundreds digits. What do you notice?

Think and Grow: Understand Place Value

A **place value chart** shows the value of each digit in a number.
The value of each place is 10 times the value of the place to the right.

The place value chart shows how the place values are grouped.
Each group is called a **period**. In a number, periods are separated by commas.

Thousands Period			Ones Period		
Hundreds	**Tens**	**Ones**	**Hundreds**	**Tens**	**Ones**
4	2	7,	6	8	2

4 hundred thousands	2 ten thousands	7 thousands	6 hundreds	8 tens	2 ones
400,000	20,000	7,000	600	80	2

Example

Thousands Period			Ones Period		
Hundreds	**Tens**	**Ones**	**Hundreds**	**Tens**	**Ones**
2	7	5,	4	4	9

- The number, in standard form, is ________.
- The value of the digit 7 is 7 ten thousands, or ________.
- The value of the digit 4 in the hundreds place is ________.
- The value of the digit 4 in the tens place is ________.
- The value of the digit 4 in the hundreds place is ________ times the value of the digit 4 in the tens place.

Show and Grow *I can do it!*

Write the value of the underlined digit.

1. 93,<u>5</u>17

2. 6<u>8</u>5,726

3. 35<u>9</u>,842

4. <u>4</u>83,701

Compare the values of the underlined digits.

5. <u>2</u>0 and <u>2</u>00

6. <u>1</u>,000 and <u>1</u>00

Name ______________________________

Apply and Grow: Practice

Write the value of the underlined digit.

7. $45,80\underline{2}$	**8.** $9\underline{7},361$	**9.** $168,3\underline{9}2$	**10.** $\underline{8}07,516$
11. $400,\underline{5}32$	**12.** $7\underline{4}9,263$	**13.** $\underline{6}19,457$	**14.** $30\underline{1},882$

Compare the values of the underlined digits.

15. $\underline{4}$ and $\underline{4}7$

16. $\underline{3}5,649$ and $2\underline{3},799$

17. Your friend is 9 years old. Your neighbor is 90 years old. Your neighbor is how many times as old as your friend?

18. **YOU BE THE TEACHER** Is Newton correct? Explain.

19. **Writing** Explain the relationship between the place values when the same two digits are next to each other in a multi-digit number.

Think and Grow: Modeling Real Life

Example What is the value of the digit 3 in the distance around Saturn? in the distance around Jupiter? How do these values relate to each other?

Planet	Distance Around Planet (kilometers)
Mercury	15,329
Venus	38,025
Earth	40,030
Mars	21,297
Jupiter	439,264
Saturn	365,882
Uranus	159,354
Neptune	154,705

Write each number in a place value chart.

	Thousands Period			Ones Period		
	Hundreds	Tens	Ones	Hundreds	Tens	Ones
Saturn						
Jupiter						

Compare the values of the digit 3 in each number.

Saturn: The value of the digit 3 is ________.

Jupiter: The value of the digit 3 is ________.

The value of the digit 3 in the distance around Saturn is _____ times the value of the digit 3 in the distance around Jupiter.

Show and Grow *I can think deeper!*

Use the table above.

20. The distance around which planet has an 8 in the thousands place?

21. Compare the value of the 8s in the distance around Saturn.

22. What is the value of the digit 4 in the distance around Earth? in the distance around Neptune? How do these values relate to each other?

23. Compare the values of the first digits in the distances around Earth and Jupiter. Explain how you can use the values to compare the sizes of these two planets.

Name ____________________

Homework & Practice 1.1

Learning Target: Identify the values of digits in multi-digit numbers.

Example

Thousands Period			Ones Period		
Hundreds	**Tens**	**Ones**	**Hundreds**	**Tens**	**Ones**
8	3	9,	9	5	2

- The number, in standard form, is 839,952 .
- The value of the digit 8 is 8 hundred thousands or 800,000 .
- The value of the digit 9 in the thousands place is 9,000 .
- The value of the digit 9 in the hundreds place is 900 .
- The value of the digit 9 in the thousands place is 10 times the value of the digit 9 in the hundreds place.

Write the value of the underlined digit.

1. $79{,}0\underline{4}3$

2. $52{,}61\underline{8}$

3. $3\underline{7}9{,}021$

4. $958{,}\underline{6}41$

5. $\underline{2}03{,}557$

6. $14\underline{5}{,}860$

7. $497{,}\underline{3}84$

8. $6\underline{1}2{,}739$

Compare the values of the underlined digits.

9. $\underline{6}03$ and $\underline{6}{,}425$

10. $\underline{9}30{,}157$ and $8\underline{9}{,}216$

11. A car can travel 50 miles per hour. A tsunami can travel 500 miles per hour. The tsunami is how many times faster than the car?

12. MP **Number Sense** In the number 93,825, is the value in the ten thousands place 10 times the value in the thousands place? Explain.

13. MP **Reasoning** Write the greatest number possible using each number card once. Then write the least six-digit number possible.

6 1 3 8 9 5

Greatest: ______ Least: ______

Modeling Real Life Use the table.

14. The height of which mountain has a 3 in the thousands place?

15. What is the value of the digit 5 in the height of K2? in the height of Mount Everest? How do these values relate to each other?

16. DIG DEEPER! The tallest mountain in the world is shown in the table. Which mountain is it?

Mountain	Height (feet)
K2	28,251
Mont Blanc	15,771
Mount Everest	29,035
Mount Kinabalu	13,455
Mount Rainier	14,411
Mount Whitney	14,494

Review & Refresh

17. Use the graph to answer the questions.

How many seconds did the Peach Street traffic light stay red?

How many more seconds did the Valley Road traffic light stay red than the Elm Street traffic light?

Time Traffic Lights Stay Red	
Elm Street	● ●
3rd Street	● ● ● ●
Peach Street	● ● ● ● ● ◖
Valley Road	● ● ● ◖

Each ● = 10 seconds.

Name ____________________

Read and Write Multi-Digit Numbers 1.2

Learning Target: Read and write multi-digit numbers in different forms.

Success Criteria:

- I can write a number in standard form.
- I can read and write a number in word form.
- I can write a number in expanded form.

Model each number. Draw to show your models. Then write each number a different way.

2,186

five thousand, two hundred thirteen

3,000 + 600 + 90 + 4

Structure Model a different four-digit number. Write the number as many different ways as possible. Compare your work to your partner's.

Think and Grow: Read and Write Multi-Digit Numbers

Example Write the number in standard form, word form, and expanded form.

Thousands Period			Ones Period		
Hundreds	Tens	Ones	Hundreds	Tens	Ones
4	2	7,	5	6	1

Standard form:

Word form:

Expanded form: ________ + ________ + ________ + ______ + _____ + _____

Example Use the place value chart to write the number below in standard form and expanded form.

Thousands Period			Ones Period		
Hundreds	Tens	Ones	Hundreds	Tens	Ones

Standard form:

Word form: fifty-four thousand, two

Expanded form: ________ + ________ + _____

Show and Grow *I can do it!*

Write the number in two other forms.

1. Standard form: 38,650

Word form:

Expanded form:

2. Standard form:

Word form: one hundred five thousand, ninety-eight

Expanded form:

Name ______

Apply and Grow: Practice

Write the number in two other forms.

3. Standard form:

Word form:

Expanded form: 600,000 + 40,000 + 2,000 + 50

4. Standard form: 134,078

Word form:

Expanded form:

5. Complete the table.

Standard Form	Word Form	Expanded Form
	three thousand, four hundred ninety-seven	
		50,000 + 2,000 + 400 + 80
	eighty-eight thousand, one hundred six	
610,010		

6. **YOU BE THE TEACHER** The fangtooth fish lives about 6,500 feet underwater. Newton reads this number as "six thousand five hundred." Descartes reads the number as "sixty-five hundred." Did they both read the number correctly? Explain.

7. **Which One Doesn't Belong?** Which one does *not* belong with the other three?

eight hundred sixteen thousand, nine

800,000 + 10,000 + 6,000 + 900

816,009

Thousands Period			Ones Period		
Hundreds	Tens	Ones	Hundreds	Tens	Ones
8	1	6,	0	0	9

Think and Grow: Modeling Real Life

The *telegraph* was the machine used to transmit messages written in Morse code.

Example Morse code is a code in which numbers and letters are represented by a series of dots and dashes. Use the table to write the number in standard form, word form, and expanded form.

•••–– ––––• ••••– ––––– –––•• •––––

Standard form:

Word form:

Expanded form:

Morse Code Numbers			
0	–––––	5	•••••
1	•––––	6	–••••
2	••–––	7	––•••
3	•••––	8	–––••
4	••••–	9	––––•

Show and Grow *I can think deeper!*

8. Use the table above to write the number in standard form, word form, and expanded form.

––••• ••••• ––––– ••––– ––––– –••••

9. Use the number 20,000 + 9,000 + 400 + 50 to complete the check.

0000

DATE

PAY TO THE ORDER OF *Compassionate Care for All Animals* $ [ ] ← standard form

word form → ______ DOLLARS

MEMO *Donation*

AUTHORIZED SIGNATURE

123456789 09 02 12346694566 0003

Name ______________________________

Learning Target: Read and write multi-digit numbers in different forms.

Example Use the place value chart to write the number below in word form and expanded form.

Thousands Period			Ones Period		
Hundreds	**Tens**	**Ones**	**Hundreds**	**Tens**	**Ones**
5	6	4,	0	8	7

Standard form: 564,087

Word form: five hundred sixty-four thousand, eighty-seven

Expanded form: 500,000 + 60,000 + 4,000 + 80 + 7

Write the number in two other forms.

1. Standard form:

Word form:

Expanded form: 500,000 + 40,000 + 3,000 + 200 + 90 + 8

2. Standard form:

Word form: forty-eight thousand, six hundred three

Expanded form:

3. Complete the table.

Standard Form	Word Form	Expanded Form
9,629		
		30,000 + 7,000 + 800 + 2
100,002		
	four hundred sixteen thousand, eighty-seven	

4. **MP Reasoning** Your teacher asks the class to write forty-two thousand, ninety-three in standard form. Which student wrote the correct number? What mistake did the other student make?

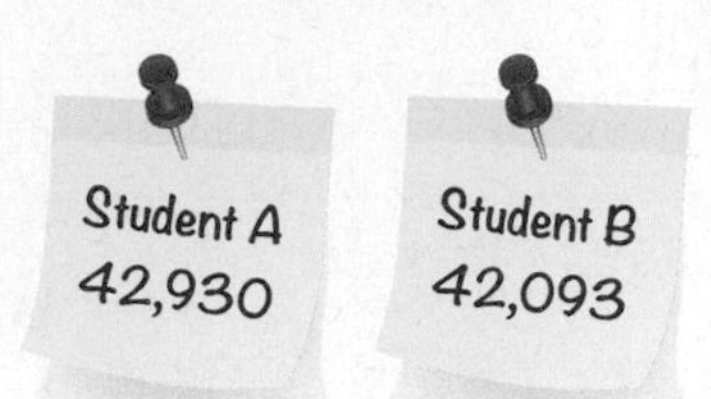

5. **MP Logic** What is the number?

The number has two periods. The thousands period is written as six hundred eight thousand in word form. The ones period is written as 600 + 80 in expanded form.

6. **Modeling Real Life** Use the table to write the number in standard form, word form, and expanded form.

Braille Numbers

1	2	3	4	5	6	7	8	9	0

7. **Modeling Real Life** Use the number 3,000 + 70 + 1 to complete the check.

Review & Refresh

Compare.

8. $\frac{1}{6} \bigcirc \frac{2}{6}$

9. $\frac{2}{2} \bigcirc \frac{2}{3}$

10. $\frac{1}{2} \bigcirc \frac{3}{4}$

11. $\frac{1}{4} \bigcirc \frac{2}{8}$

Name ____________________

Compare Multi-Digit Numbers 1.3

Learning Target: Use place value to compare two multi-digit numbers.

Success Criteria:

- I can explain how to compare two numbers with the same number of digits.
- I can use the symbols <, >, and = to compare two numbers.

Goal: Make the greatest number possible.

Draw a Number Card. Choose a place value for the digit. Write the digit in the place value chart. Continue until the place value chart is complete.

Thousands Period			Ones Period		
Hundreds	Tens	Ones	Hundreds	Tens	Ones

Compare your number with your partner's. Whose number is greater?

Construct Arguments Explain your strategy to your partner. Compare your strategies.

Think and Grow: Compare Multi-Digit Numbers

60 > 30

Example Compare 8,465 and 8,439.

Use a place value chart.

Thousands Period			Ones Period		
Hundreds	**Tens**	**Ones**	**Hundreds**	**Tens**	**Ones**
		8,	4	6	5
		8,	4	3	9

Start at the left. Compare the digits in each place until the digits differ.

Step 1: Compare the thousands.

8,465
8,439

8 thousands (=) 8 thousands

Step 2: Compare the hundreds.

8,465
8,439

4 hundreds (=) 4 hundreds

Step 3: Compare the tens.

8,465
8,439

6 tens (>) 3 tens

So, 8,465 (>) 8,439.

Show and Grow *I can do it!*

Write which place to use when comparing the numbers.

1. 2,423 / 2,324 hundreds
2. 9,631 / 9,637 ones
3. 15,728 / 16,728 thousands

Compare.

4. 5,049 (<) 5,082
5. 735,283 (=) 735,283 
6. 43,694 (>) 3,694
7. 88,195 (>) 78,195
8. 6,480 (<) 6,508
9. 321,817 (<) 312,827 

Name ____________________

Apply and Grow: Practice

Compare.

10. 6,052 ○ 6,520

11. 891,634 ○ 871,634

12. 28,251 ○ 26,660

13. 324,581 ○ 32,458

14. 230,611 ○ 230,610

15. 909,900 ○ 909,009

16. 7,000 + 100 + 30 + 4 ○ 7,634

17. 100,003 ○ ten thousand, three

18. sixteen thousand, four hundred nine ○ 16,490

19. 400,000 + 60,000 + 1,000 + 300 + 20 + 9 ○ 461,329

20. Two brands of televisions cost $1,598 and $1,998. Which is the lesser price?

21. **DIG DEEPER!** Your friend says she can tell which sum is greater without adding the numbers. How can she tell?

34,593 + 6,781

34,593 + 6,609

22. **MP Number Sense** Write all of the digits that make the number greater than 23,489 and less than 26,472.

2 ___?___ ,650

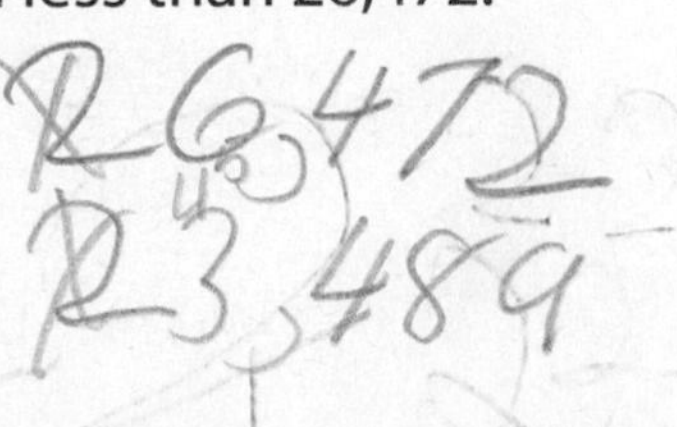

23. **YOU BE THE TEACHER** Your friend says 38,675 is less than 9,100 because 3 is less than 9. Is your friend correct? Explain.

Think and Grow: Modeling Real Life

Example Which playground costs the least?

Write each number in a place value chart.

	Thousands Period			Ones Period		
	Hundreds	**Tens**	**Ones**	**Hundreds**	**Tens**	**Ones**
Yellow						
Green						
Blue						

Playground Prices	
Yellow	$36,827
Green	$35,872
Blue	$36,927

Order the numbers from least to greatest.

________, ________, ________

The ________ playground costs the least.

Show and Grow *I can think deeper!*

24. Who received the highest score? Did anyone beat the high score?

Pinball Scores	
You	254,020
Friend	245,140
Cousin	245,190

25. Name two cities that have a greater population than St. Louis. Name two cities that do not have a greater population than Oakland.

U.S. Census City Populations	
St. Louis, MO	319,294
Tulsa, OK	391,906
Anchorage, AK	291,826
Cleveland, OH	396,815
Oakland, CA	390,724

Name ______________________________

1.4 Round Multi-Digit Numbers

Learning Target: Use place value to round multi-digit numbers.

Success Criteria:

- I can explain which digit I use to round and why.
- I can round a multi-digit number to any place.

Explore and Grow

Write five numbers that round to 250 when rounded to the nearest ten.

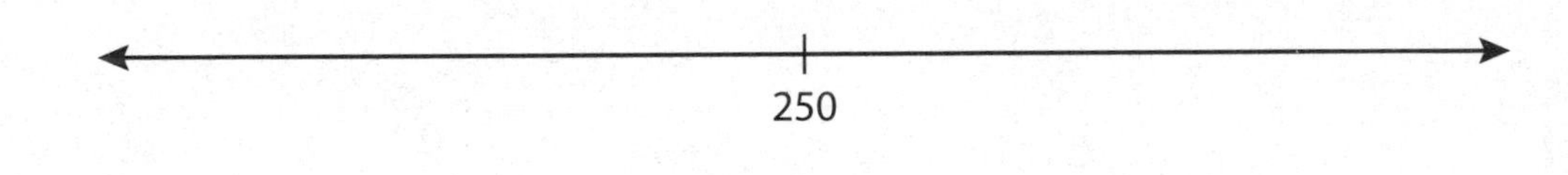

_____ _____ _____ _____ _____

Write five numbers that round to 500 when rounded to the nearest hundred.

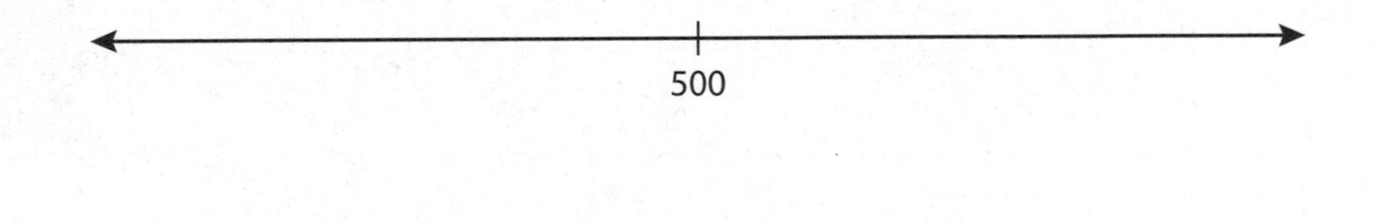

_____ _____ _____ _____ _____

MP **Repeated Reasoning** Explain your strategy. Then explain how you could use the strategy to round a four-digit number to the nearest thousand.

Think and Grow: Round Multi-Digit Numbers

To round a number, find the multiple of 10, 100, 1,000, and so on, that is closest to the number. You can use a number line or place value to round numbers.

Example Use a number line to round 4,276 to the nearest thousand.

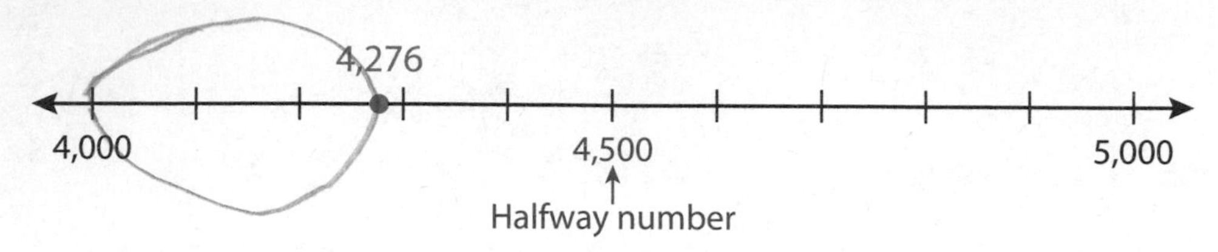

4,276 is closer to 4,000 than it is to 5,000.

So, 4,276 rounded to the nearest thousand is ________.

Example Use place value to round 385,617 to the nearest ten thousand.

Remember, if the digit to the right of the rounding digit is 5 or greater, then the rounding digit increases by one.

ten thousands place   5 = 5

3 8 5,6 1 7
↓ ↓ ↓ ↓ ↓ ↓
3 9 0,0 0 0

So, 385,617 rounded to the nearest ten thousand is ________.

Show and Grow *I can do it!*

Round the number to the place of the underlined digit.

1. 6,912

2. 43,215

3. 25,883

4. 148,796

5. Round 5,379 to the nearest thousand.

6. Round 70,628 to the nearest ten thousand.

7. Round 362,113 to the nearest hundred thousand.

8. Round 982,638 to the nearest thousand.

Name ______________________________

Apply and Grow: Practice

Round the number to the place of the underlined digit.

9. $3,\underline{6}41$

10. $1\underline{7},139$

11. $426,\underline{3}84$

12. $54\underline{2},930$

Round the number to the nearest thousand.

13. 9,426

14. 57,496

15. 360,491

16. 824,137

Round the number to the nearest hundred thousand.

17. 226,568

18. 457,724

19. 108,665

20. 75,291

21. MP **Number Sense** Write a five-digit number that has the digits 6, 0, 4, 2, and 8 and rounds to 70,000 when rounded to the nearest ten thousand.

22. MP **Number Sense** When finding the United States census, should you round or find an exact answer? Explain.

23. **Open-Ended** A lightning strike can reach a temperature of about 54,000 degrees Fahrenheit. Write four possible temperatures for a lightning strike.

Think and Grow: Modeling Real Life

Example When the results are rounded to the nearest ten thousand, which token received about 160,000 votes?

Round each number to the nearest ten thousand.

134,704 → ______ 167,582 → ______

154,165 → ______ 146,661 → ______

207,954 → ______ 165,083 → ______

212,476 → ______ 160,485 → ______

So, the ____________ received about 160,000 votes.

Favorite MONOPOLY® Tokens	
Token	**Number of Votes**
Battleship	134,704
Cat	154,165
Dinosaur	207,954
Dog	212,476
Hat	167,582
Penguin	146,661
Race Car	165,083
Rubber Duck	160,485

Show and Grow *I can think deeper!*

Use the table above.

24. When the results are rounded to the nearest ten thousand, which tokens received about 150,000 votes?

25. When the results are rounded to the nearest hundred thousand, which tokens received about 100,000 votes?

26. **DIG DEEPER!** To which place should you round each number of votes so that you can order the numbers from greatest to least? Explain.

27. A car battery should be replaced when the odometer shows about 50,000 miles. None of the cars below have had a battery replacement. Which cars might need a new battery?

ODO
5,617 mi

ODO
44,534 mi

ODO
53,798 mi

ODO
49,142 mi

ODO
35,961mi

Name ______________________________

Performance Task 1

You hike from Point A through Point F along the orange path shown on the map.

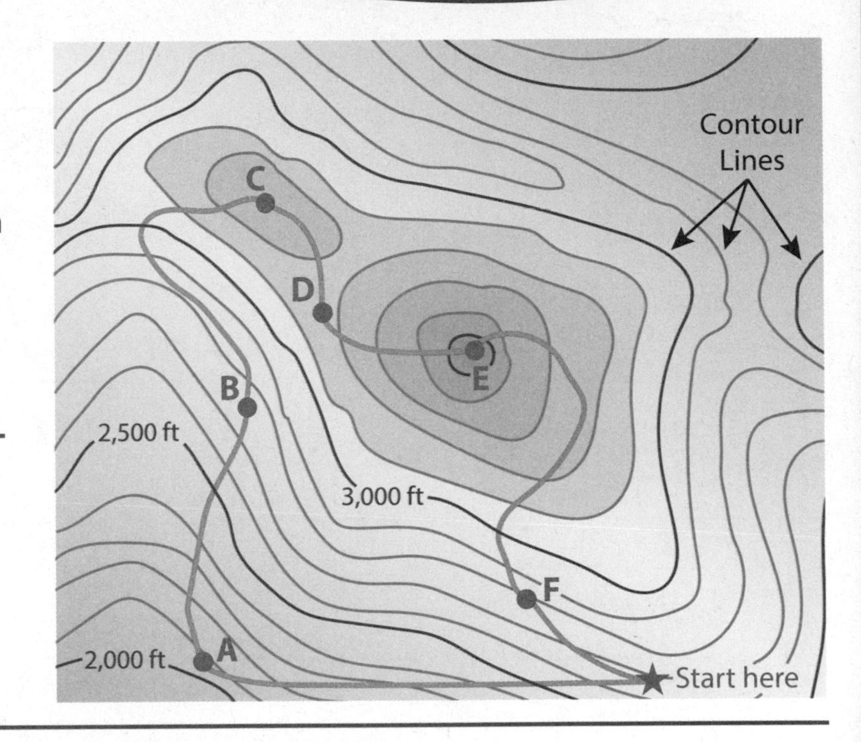

1. What is the distance in elevation between each contour line?

____

2. As you walk from A to C, are you walking uphill or downhill? Explain.

____

3. Which letter represents the highest point? Estimate the height.

____

4. A water station has an elevation of 2,763 feet. Which letter represents the location of the water station?

____

5. You take a break when you are at two thousand eighty feet. At which letter do you take a break?

____

6. About how much higher do you think Point C is than Point D? Explain.

Place Value Plug In

Directions:

1. Players take turns.
2. On your turn, roll six dice. Arrange the dice into a six-digit number that matches one of the descriptions.
3. Write your number on the lines.
4. The first player to complete all of the numbers wins!

A number with...	Number
6 in the tens place 2 in the ten thousands place	___ ___ ___ , ___ ___ ___
4 in the ones place 5 in the thousands place	___ ___ ___ , ___ ___ ___
3 in the tens place 1 in the hundreds place	___ ___ ___ , ___ ___ ___
5 in the ones place 2 in the hundred thousands place	___ ___ ___ , ___ ___ ___
4 in the thousands place 6 in the ten thousands place	___ ___ ___ , ___ ___ ___
three of the same digit	___ ___ ___ , ___ ___ ___
digits in the hundred thousands place and ones place have a sum of 8	___ ___ ___ , ___ ___ ___
digits in the hundreds place and ten thousands place have a sum of 7	___ ___ ___ , ___ ___ ___
FREEBIE! Use any number!	___ ___ ___ , ___ ___ ___

Name ______________________________

Chapter Practice

1.1 Understand Place Value

Write the value of the underlined digit.

1. 26,490
2. 57,811
3. 308,974
4. 748,612

Compare the values of the underlined digits.

5. 94 and 982
6. 817,953 and 84,006

1.2 Read and Write Multi-Digit Numbers

7. Complete the table.

Standard Form	Word Form	Expanded Form
		50,000 + 600 + 90 + 1
	seven hundred two thousand, five hundred	
993,004		

8. **Modeling Real Life** Use the number 10,000 + 4,000 + 300 + 90 + 9 to complete the check.

1.3 Compare Multi-Digit Numbers

Write which place to use when comparing the numbers.

9. 46,027
46,029

10. 548,003
545,003

11. 619,925
630,982

Compare.

12. 4,021 ◯ 4,210

13. 78,614 ◯ 78,614

14. 816,532 ◯ 816,332

15. 55,002 ◯ 65,002

16. 3,276 ◯ 3,275

17. 45,713 ◯ 457,130

18. 569,021 ◯ 50,000 + 6,000 + 900 + 2

19. thirty-seven thousand ◯ 307,000

20. Two different hot tubs cost $4,179 and $4,139. Which is the lesser price?

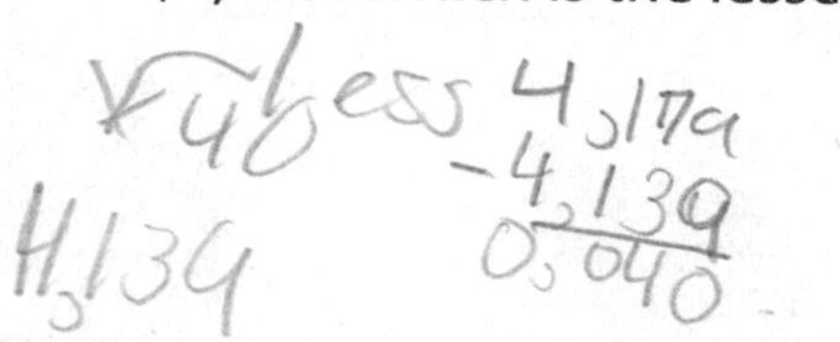

21. MP **Number Sense** Write all of the digits that make the number greater than 47,068 and less than 47,468.

47,___?___68

1.4 Round Multi-Digit Numbers

Round the number to the place of the underlined digit.

22. 8,614

23. 2,725

24. 27,602

25. 906,154

Round the number to the nearest thousand.

26. 1,358

27. 57,094

Round the number to the nearest ten thousand.

28. 431,849

29. 60,995

2 Add and Subtract Multi-Digit Numbers

- The population of an area is the number of people who live there. Why would it be important to know the population of a city?
- About how many people do you think live in your city? Why is the population usually an estimate?

Chapter Learning Target:
Understand adding and subtracting numbers.

Chapter Success Criteria:
- I can estimate a sum or difference.
- I can explain which strategy I used to write a sum or difference.
- I can write a sum or difference.
- I can solve addition and subtraction problems.

Name ______________________________

2 Vocabulary

Review Words

addends
Commutative Property of Addition
sum

Organize It

Use the review words to complete the graphic organizer.

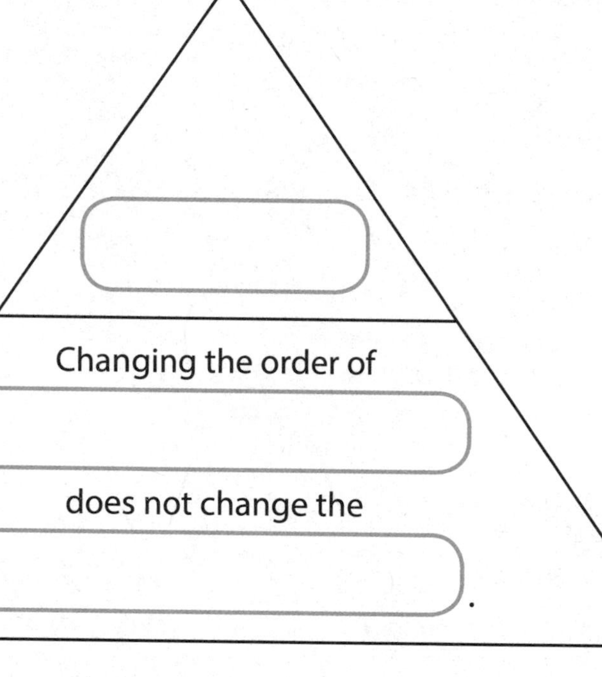

$8 + 7 = 15$ $7 + 8 = 15$

So, $8 + 7 = 7 + 8$.

Define It

Use your vocabulary card to complete the definition.

1. estimate: A ______________ that is ______________ to an ______________ number

Chapter 2 Vocabulary Cards

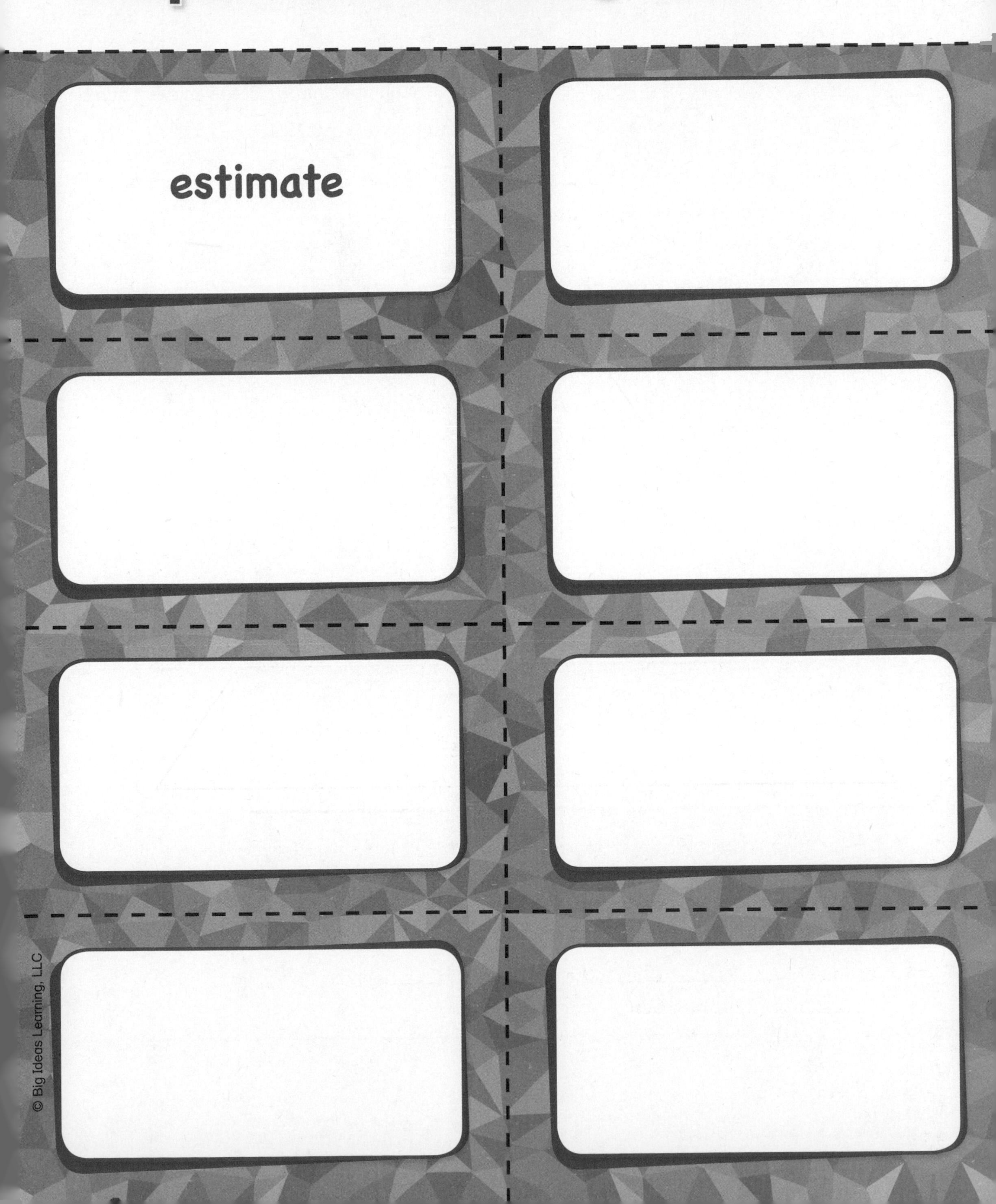

A number that is close to an exact number

8,195 + 9,726 = ?

Exact Sum: 17,921

Estimate: 18,000

Name ____________________

Estimate Sums and Differences 2.1

Learning Target: Use rounding to estimate sums and differences.

Success Criteria:
- I can use rounding to estimate a sum.
- I can use rounding to estimate a difference.
- I can explain what happens when I round to different place values.

Estimate to find each sum by rounding to the nearest thousand, hundred, or ten. Explain why you chose to round to that place value.

A football team had 917 spectators at their first game and 872 at their second game. About how many spectators did the team have at both games?

A company budgets $1,800 for a company picnic. They spend $917 on the location and $872 on food. Did they stay within their budget?

Reasoning Explain why you may choose to round to different place values in different situations.

Think and Grow: Estimate Sums and Differences

An **estimate** is a number that is close to an exact number. You can use rounding to estimate sums and differences.

Example Estimate 8,675 + 3,214.

One Way: Round each addend to the nearest hundred. Then find the sum.

8,675 → ______
+ 3,214 → + ______
______

So, 8,675 + 3,214 is about ______.

Another Way: Round each addend to the nearest thousand. Then find the sum.

8,675 → ______
+ 3,214 → + ______
______

So, 8,675 + 3,214 is about ______.

Example Estimate 827,615 − 54,306.

One Way: Round each number to the nearest thousand. Then find the difference.

827,615 → ______
− 54,306 → − ______
______

So, 827,615 − 54,306 is about ______.

Another Way: Round each number to the nearest ten thousand. Then find the difference.

827,615 → ______
− 54,306 → − ______
______

So, 827,615 − 54,306 is about ______.

Show and Grow *I can do it!*

Estimate the sum or difference.

1. 63,851 → ______
+ 19,375 → + ______
______

2. 4,874 → ______
− 2,530 → − ______
______

Name ______

Apply and Grow: Practice

Estimate the sum or difference.

3. 27,369 → ☐
+ 14,608 → + ☐
☐

4. 53,744 → ☐
− 41,086 → − ☐
☐

5. 68,451
− 40,695

6. 34,685
+ 27,043

7. 908,465
− 653,299

8. 478,633
+ 200,081

9. 395,408
− 102,677

10. 563,427
+ 178,023

11. 888,056 − 423,985 = ______

12. 713,642 + 49,018 = ______

13. **DIG DEEPER!** Is 20,549 + 9,562 greater than or less than 30,000? Explain how you know without finding the exact sum.

14. **Writing** Describe a real-life situation in which estimation would *not* be appropriate to use.

Think and Grow: Modeling Real Life

Example About how many more pounds does the whale shark weigh than the orca?

Whale Shark: 40,364 pounds

Orca: 8,095 pounds

Round the weight of each animal to the nearest thousand because you do not need a precise answer.

Orca: ________ Whale shark: ________

Subtract the estimated weight of the orca from the estimated weight of the whale shark.

$$\begin{array}{r} 40{,}364 \\ -\ 8{,}095 \\ \hline \end{array} \quad \begin{array}{c} \rightarrow \\ \rightarrow \end{array} \quad \begin{array}{r} \square \\ -\ \square \\ \hline \square \end{array}$$

The whale shark weighs about ________ more pounds than the orca.

Show and Grow *I can think deeper!*

Local Election Results	
Candidate	**Number of Votes**
Candidate A	250,311
Candidate B	84,916

15. About how many more votes did Candidate A receive than Candidate B?

16. Mount Saint Helens is a volcano that is 8,363 feet tall. Mount Fuji is a volcano that is 4,025 feet taller than Mount Saint Helens. About how tall is Mount Fuji?

17. An educational video has 6,129 fewer views than a gaming video. The educational video has 483,056 views. About how many views does the gaming video have?

Name ____________________

Homework & Practice 2.1

Learning Target: Use rounding to estimate sums and differences.

Example Estimate 871,432 − 406,895.

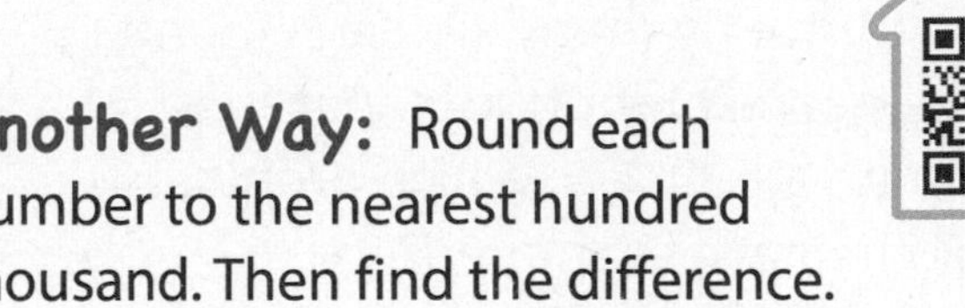

One Way: Round each number to the nearest ten thousand. Then find the difference.

871,432 → 870,000
− 406,895 → − 410,000
460,000

So, 871,432 − 406,895 is about 460,000.

Another Way: Round each number to the nearest hundred thousand. Then find the difference.

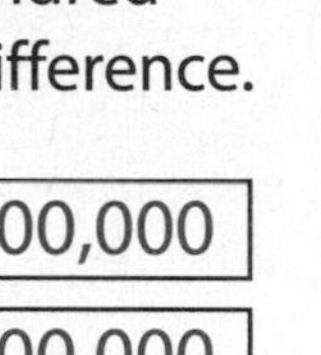

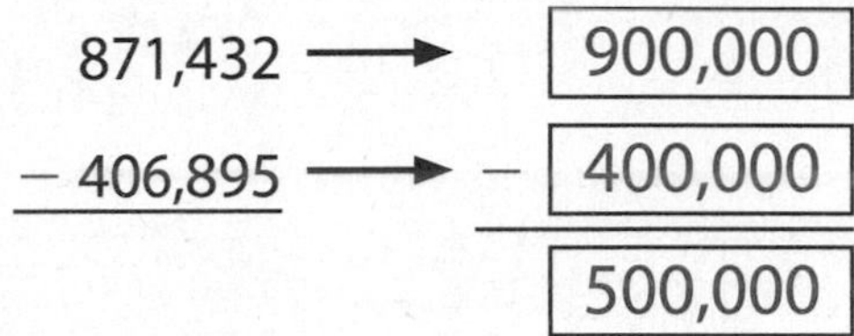

871,432 → 900,000
− 406,895 → − 400,000
500,000

So, 871,432 − 406,895 is about 500,000.

Estimate the sum or difference.

1. 7,910 → ☐
+ 1,358 → + ☐
☐

2. 5,608 → ☐
− 3,217 → − ☐
☐

3. 73,406
− 45,699

4. 82,908
+ 28,643

5. 96,420
− 63,877

6. 517,605
+ 359,421

7. 688,203
− 444,387

8. 261,586
+ 116,934

Estimate the sum or difference.

9. 864,733 − 399,608 = ________

10. 134,034 + 26,987 = ________

11. MP **Number Sense** Descartes estimates a difference by rounding each number to the nearest ten thousand. His estimate is 620,000. Which problems could he have estimated?

694,506 − 73,421	886,789 − 265,064
675,896 − 51,309	704,322 − 82,156

12. MP **Reasoning** When might you estimate the difference of 603,476 and 335,291 to the nearest hundred? to the nearest hundred thousand?

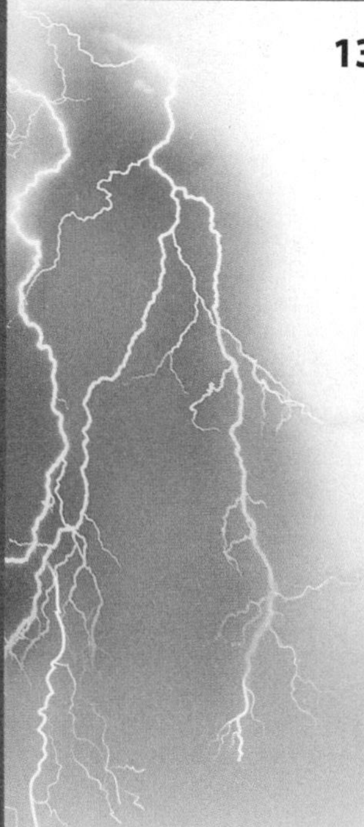

13. **Modeling Real Life** A storm causes 23,890 homes to be without power on the east side of a city and 18,370 homes to be without power on the west side. About how many homes altogether are without power?

14. **Modeling Real Life** You walk 5,682 steps. Your teacher walks 4,219 steps more than you. About how many steps does your teacher walk?

Review & Refresh

Find the product.

15. $3 \times 3 \times 2 =$ _____

16. $2 \times 4 \times 7 =$ _____

17. $3 \times 3 \times 5 =$ _____

18. $6 \times 2 \times 3 =$ _____

19. $4 \times 9 \times 2 =$ _____

20. $4 \times 10 \times 2 =$ _____

Name ______________________________

Add Multi-Digit Numbers 2.2

Learning Target: Add multi-digit numbers and check whether the sum is reasonable.

Success Criteria:

- I can use place value to line up the numbers in an addition problem.
- I can add multi-digit numbers, regrouping when needed.
- I can estimate a sum to check whether my answer is reasonable.

Explore and Grow

Which addition problem shows a correct way to find 38 + 7? Why?

$$\begin{array}{r} 38 \\ +7\ \ \\ \hline \end{array} \qquad \begin{array}{r} 38 \\ +\ \ 7 \\ \hline \end{array}$$

Which addition problem shows a correct way to find 403 + 1,248? Why?

$$\begin{array}{r} 403 \\ +\ 1,248 \\ \hline \end{array} \qquad \begin{array}{r} 1,248 \\ +\ 403\ \ \\ \hline \end{array}$$

Reasoning Why do you need to use place value when adding? Explain.

Think and Grow: Add Multi-Digit Numbers

Example Add: 307,478 + 95,061.

Estimate: 307,000 + 95,000 = ________

Use place value to line up the addends.

$$\begin{array}{r} 307{,}478 \\ +\ \ 95{,}061 \\ \hline \end{array}$$

Add the ones, then the tens, and then the hundreds. Regroup if necessary.

$$\begin{array}{r} 1 \\ 307{,}478 \\ +\ \ 95{,}061 \\ \hline 539 \end{array}$$

Regroup 13 tens.

Add the thousands, then the ten thousands, and then the hundred thousands. Regroup if necessary.

$$\begin{array}{r} 1\,1\ \ 1 \\ 307{,}478 \\ +\ \ 95{,}061 \\ \hline 402{,}539 \end{array}$$

Regroup 12 thousands.

Regroup 10 ten thousands.

Regroup 10 ten thousands as 1 hundred thousand and 0 ten thousands.

The sum is 402,539.

Check: Because ________ is close to the estimate, ________, the answer is reasonable.

Show and Grow I can do it!

Find the sum. Check whether your answer is reasonable.

1. Estimate: ________

$$\begin{array}{r} 17{,}690 \\ +\ 53{,}024 \\ \hline \end{array}$$

2. Estimate: ________

$$\begin{array}{r} 297{,}853 \\ +\ \ \ \ 6{,}129 \\ \hline \end{array}$$

Name ______________________________

Apply and Grow: Practice

Find the sum. Check whether your answer is reasonable.

3. Estimate: ________

$$\begin{array}{r} 6{,}439 \\ +\ 3{,}278 \\ \hline \end{array}$$

4. Estimate: ________

$$\begin{array}{r} 61{,}096 \\ +\ 8{,}765 \\ \hline \end{array}$$

5. Estimate: ________

$$\begin{array}{r} 82{,}238 \\ +\ 4{,}697 \\ \hline \end{array}$$

6. Estimate: ________

$$\begin{array}{r} 46{,}792 \\ +\ 38{,}516 \\ \hline \end{array}$$

7. Estimate: ________

$$\begin{array}{r} 686{,}420 \\ +\ 75{,}319 \\ \hline \end{array}$$

8. Estimate: ________

$$\begin{array}{r} 594{,}341 \\ +\ 307{,}899 \\ \hline \end{array}$$

9. Estimate: ________

246,890 + 13,579 = ________

10. Estimate: ________

822,450 + 8,651 = ________

11. A video receives 10,678 views the first day. It receives 25,932 views the second day. How many views does the video receive in two days?

12. **YOU BE THE TEACHER** Is Newton correct? Explain.

13. **DIG DEEPER!** Find the missing digits.

$$\begin{array}{r} 7\ 5\ 9,\square\ 7\ 0 \\ +\ 2\ 3\ \square,1\ 9\ 2 \\ \hline 9\ 9\ 3,7\ 6\ 2 \end{array}$$

Think and Grow: Modeling Real Life

Example A family is traveling in a car from Seattle to Atlanta. They travel 1,099 miles the first two days and 1,082 miles the next two days. Has the family arrived in Atlanta?

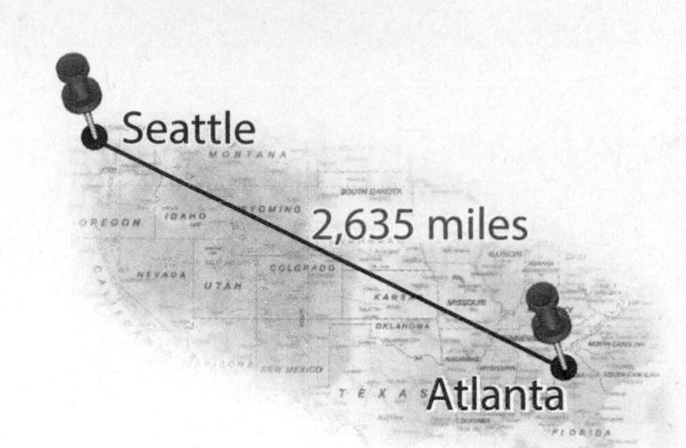

Add the distances traveled.

$$\begin{array}{r} 1{,}099 \\ +\ 1{,}082 \\ \hline \end{array}$$

Compare the distance traveled to the distance from Seattle to Atlanta.

The family ______________ arrived in Atlanta.

Show and Grow *I can think deeper!*

14. One World Trade Center has 2,226 steps. A visitor enters the building and climbs 1,387 steps, takes a break, and climbs 839 more steps. Did the visitor reach the top of the building?

15. There were 51,787 more people who rode the city buses on Saturday than on Sunday. On Sunday, 174,057 people rode the buses. How many people rode the buses on Saturday?

16. The deepest part of the Atlantic Ocean is 8,577 feet shallower than the deepest part of the Pacific Ocean. The deepest part of the Atlantic Ocean is 27,493 feet deep. How deep is the deepest part of the Pacific Ocean?

Name ______________________

Subtract Multi-Digit Numbers 2.3

Learning Target: Subtract multi-digit numbers and check my answer.

Success Criteria:

- I can use place value to line up the numbers in a subtraction problem.
- I can subtract multi-digit numbers, regrouping when needed.
- I can estimate a difference or use addition to check my answer.

Explore and Grow

Which subtraction problem shows a correct way to find 94 − 8? Why?

$$\begin{array}{r} 94 \\ -\ \ 8 \\ \hline \end{array} \qquad \begin{array}{l} 94 \\ -8 \\ \hline \end{array}$$

Which subtraction problem shows a correct way to find 3,710 − 251? Why?

$$\begin{array}{l} 3{,}710 \\ -251 \\ \hline \end{array} \qquad \begin{array}{r} 3{,}710 \\ -\quad 251 \\ \hline \end{array}$$

Reasoning Why do you need to use place value when subtracting? Explain.

Think and Grow: Subtract Multi-Digit Numbers

Example Subtract: 60,751 − 8,419.

Use place value to line up the numbers.

```
  60,751
−  8,419
```

Subtract the ones, then the tens, and then the hundreds. Regroup if necessary.

```
       4 11
  6 0, 7 5 1
−   8, 4 1 9
       3 3 2
```

Regroup 5 tens and 1 one.

Subtract the thousands and then the ten thousands. Regroup if necessary.

```
  5 10   4 11
  6 0, 7 5 1
−   8, 4 1 9
  5 2, 3 3 2
```

Regroup 6 ten thousands and 0 thousands as _____ ten thousands and _____ thousands.

The difference is 52,332.

Check: Use addition to check your answer.

```
  52,332
+  8,419
```

Show and Grow *I can do it!*

Find the difference. Then check your answer.

1.
```
  9,256
− 7,183
```

2.
```
  23,846
− 14,025
```

3.
```
  844,923
− 392,816
```

4.
```
  268,501
−  85,402
```

Name ______________________________

Apply and Grow: Practice

Find the difference. Then check your answer.

5. 96,090 − 5,130

6. 42,648 − 9,169

7. 57,502 − 4,380

8. 43,629 − 18,101

9. 425,631 − 86,942

10. 600,470 − 307,281

11. 281,660 − 44,521 = ________

12. 798,400 − 5,603 = ________

13. 103,219 people attended a championship football game last year. 71,088 people attend the game this year. How many more people attended the game last year than this year?

14. MP **Number Sense** Find and explain the error. What is the correct difference?

435,450
− 71,945
444,515

15. MP **Number Sense** Which statements describe the difference of 32,064 and 14,950?

The difference is about 17,000.

The difference is less than 17,000.

The difference is greater than 17,000.

The difference is 17,000.

Think and Grow: Modeling Real Life

Example The shoreline of Lake Michigan is 1,090 miles shorter than the shoreline of Lake Superior. How long is the shoreline of Lake Michigan?

Subtract 1,090 from the length of the shoreline of Lake Superior.

$$\begin{array}{r} 2{,}730 \\ -\ 1{,}090 \\ \hline \end{array}$$

Lake Superior:
2,730 miles

Lake Michigan:
? miles

The shoreline of Lake Michigan is ________ miles long.

Show and Grow *I can think deeper!*

16. The new car is $15,760 less than the new truck. How much does the new car cost?

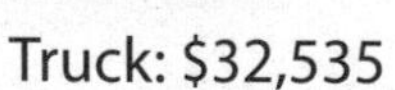

Truck: $32,535 Car: $?

17. Last year, an amusement park had 770,495 more guests than a water park. The attendance at the amusement park was 875,562 guests. What was the attendance at the water park?

18. How many fewer miles did the pilot fly in Years 1 and 2 combined than in Year 3?

Year	Miles Flown
Year 1	5,396
Year 2	10,821
Year 3	21,623

Name ______________________________

Use Strategies to Add and Subtract 2.4

Learning Target: Use strategies to add and subtract multi-digit numbers.

Success Criteria:
- I can use strategies to add multi-digit numbers.
- I can use strategies to subtract multi-digit numbers.

Choose any strategy to find 8,005 + 1,350.

Addition and Subtraction Strategies
Partial Sums
Compensation
Counting On
Regrouping

Choose any strategy to find 54,000 − 10,996.

Reasoning Explain why you chose your strategies. Compare your strategies to your partner's strategies. How are they the same or different?

Think and Grow: Use Strategies to Add or Subtract

Example Add: 3,025 + 2,160.

One Way: Use partial sums to add.

$$\begin{array}{rl} 3{,}025 = & 3{,}000 \qquad + 20 + 5 \\ +2{,}160 = & 2{,}000 + 100 + 60 \\ \hline & 5{,}000 + 100 + 80 + 5 = \_\_\_\_ \end{array}$$

Another Way: Use compensation to add.

$$\begin{array}{r} 3{,}025 - 25 \\ + 2{,}160 \\ \hline \end{array} \longrightarrow \begin{array}{r} 3{,}000 \\ + 2{,}160 \\ \hline 5{,}160 \end{array}$$

You added 25 less than 3,025, so you must add 25 to the answer.

$$\begin{array}{r} 5{,}160 \\ + \quad 25 \\ \hline \boxed{} \end{array}$$

So, 3,025 + 2,160 = ________.

Example Subtract: 16,000 − 5,984.

One Way: Count on to subtract.

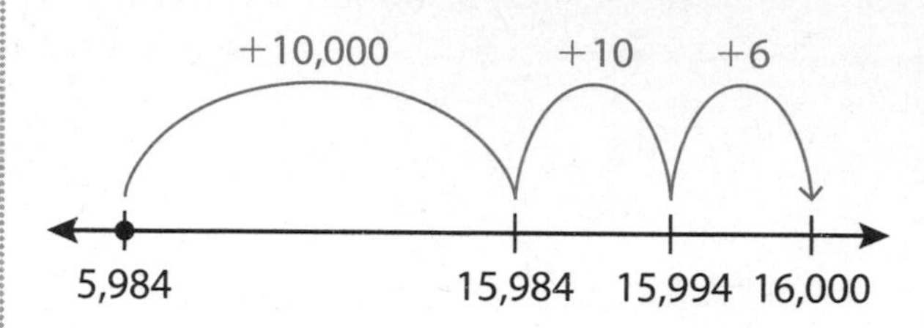

10,000 + 10 + 6 = ________

Another Way: Use compensation to subtract.

$$\begin{array}{r} 16{,}000 \\ - \ 5{,}984 + 16 \\ \hline \end{array} \longrightarrow \begin{array}{r} 16{,}000 \\ - \ 6{,}000 \\ \hline 10{,}000 \end{array}$$

You subtracted 16 more than 5,984, so you must add 16 to the answer.

$$\begin{array}{r} 10{,}000 \\ + \quad 16 \\ \hline \boxed{} \end{array}$$

So, 16,000 − 5,984 = ________.

Show and Grow *I can do it!*

Find the sum or difference. Then check your answer.

1. $\begin{array}{r} 10{,}500 \\ + 12{,}219 \\ \hline \end{array}$

2. $\begin{array}{r} 9{,}318 \\ - 7{,}008 \\ \hline \end{array}$

Name ______________________________

Apply and Grow: Practice

Find the sum or difference. Then check your answer.

3. 12,982 + 3,475

4. 44,561 − 8,480

5. 91,803 − 80,740

6. 87,871 + 13,102

7. 246,132 + 35,400

8. 860,709 − 38,115

9. 780,649 − 13,754 = ________

10. 417,890 + 90,284 = ________

11. 614,008 + 283,192 = ________

12. 801,640 − 206,427 = ________

13. MP **Structure** Write an equation shown by the number line.

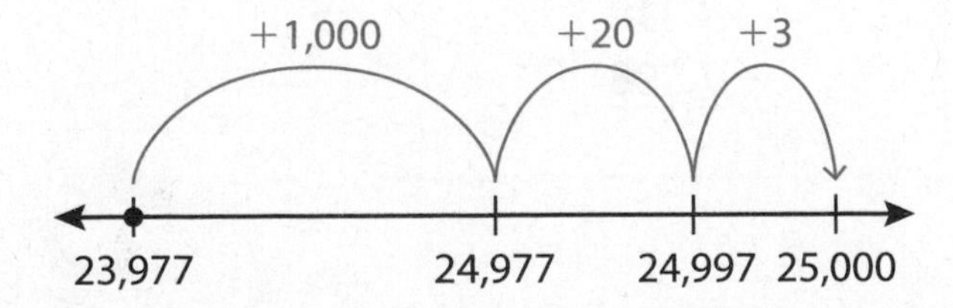

Think and Grow: Modeling Real Life

Example Earth's mantle is 1,802 miles thick. Earth's outer core is 1,367 miles thick. How many miles thinner is Earth's outer core than its mantle?

Inner Core
Outer Core
Mantle
Crust

Subtract the thickness of the outer core from the thickness of the mantle. Use compensation to subtract.

$$\begin{array}{r} 1{,}802 \\ -\underline{1{,}367}+33 \end{array} \longrightarrow \begin{array}{r} 1{,}802 \\ \underline{-\ 1{,}400} \\ 402 \end{array}$$

You subtracted 33 more than 1,367, so you must add 33 to the answer.

$$\begin{array}{r} 402 \\ \underline{+\quad 33} \\ \boxed{} \end{array}$$

Earth's outer core is ________ miles thinner than its mantle.

Show and Grow *I can think deeper!*

14. Your friend's heart beats 144,000 times in one day. Your heart beats 115,200 times in one day. How many fewer times does your heart beat than your friend's?

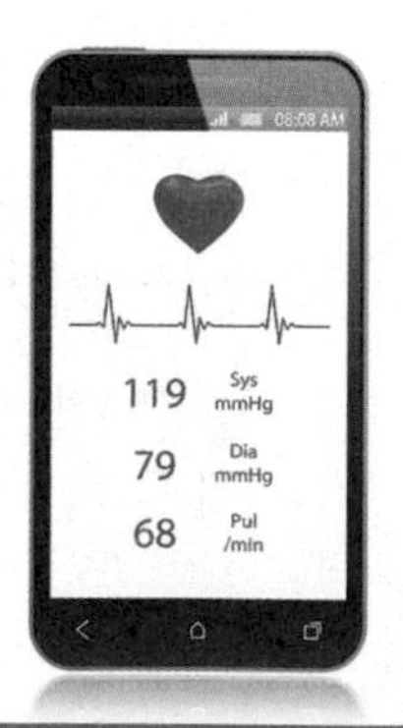

15. You, your friend, and your cousin are playing a video game. You score 2,118 more points than your friend. Your friend scores 1,503 fewer points than your cousin. What is each player's score? Who wins?

Player	Score
You	?
Friend	6,010
Cousin	?

16. Students at a school want to recycle a total of 50,000 cans and bottles. So far, the students recycled 40,118 cans and 9,863 bottles. Did the students reach their goal? If not, how many more cans and bottles need recycled?

Name ______________________________

Problem Solving: Addition and Subtraction

2.5

Learning Target: Use the problem-solving plan to solve two-step addition and subtraction word problems.

Success Criteria:

- I can understand a problem.
- I can make a plan to solve a problem using letters to represent the unknown numbers.
- I can solve a problem and check whether my answer is reasonable.

Explore and Grow

Use addition or subtraction to make a conclusion about the table.

State Land Area	
State	**Area (square miles)**
Arizona	113,594
California	155,779
Nevada	109,781
Utah	82,170

Precision Switch papers with your partner. Check your partner's answer for reasonableness.

Think and Grow: Problem Solving: Addition and Subtraction

Example You have 3,914 songs in your music library. You download 1,326 more songs. Then you delete 587 songs. How many songs do you have now?

Understand the Problem

What do you know?

- You have 3,914 songs.
- You download 1,326 more.
- You delete 587 songs.

What do you need to find?

- You need to find how many songs you have now.

Make a Plan

How will you solve?

- Add 3,914 and 1,326 to find how many songs you have after downloading some songs.
- Then subtract 587 from the sum to find how many songs you have now.

Solve

Step 1:

3,914	1,326
k	

k is the unknown sum.

$$3{,}914 + 1{,}326 = k$$

$$\begin{array}{r} 3{,}914 \\ +\ 1{,}326 \\ \hline \square \end{array}$$

k = ______

Step 2:

n	587
k = ____	

n is the unknown difference.

______ − 587 = n

$$\begin{array}{r} \square \\ -\ 587 \\ \hline \square \end{array}$$

n = ______

You have ______ songs now.

Show and Grow *I can do it!*

1. Explain how you can check whether your answer above is reasonable.

Name ______________________________

Apply and Grow: Practice

Understand the problem. What do you know? What do you need to find? Explain.

2. There are about 12,762 known ant species. There are about 10,997 known grasshopper species. The total number of known ant, grasshopper, and spider species is 67,437. How many known spider species are there?

3. A quarterback threw for 66,111 yards between 2001 and 2016. His all-time high was 5,476 yards in 1 year. In his second highest year, he threw for 5,208 yards. How many passing yards did he throw in the remaining years?

Understand the problem. Then make a plan. How will you solve? Explain.

4. There are 86,400 seconds in 1 day. On most days, a student spends 28,800 seconds sleeping and 28,500 seconds in school. How many seconds are students awake, but *not* in school?

5. A pair of rhinoceroses weigh 14,860 pounds together. The female weighs 7,206 pounds. How much more does the male weigh than the female?

6. Earth is 24,873 miles around. If a person's blood vessels were laid out in a line, they would be able to circle Earth two times, plus 10,254 more miles. How many miles long are a person's blood vessels when laid out in a line?

7. Alaska has 22,041 more miles of shoreline than Florida and California combined. Alaska has 33,904 miles of shoreline. Florida has 8,436 miles of shoreline. How many miles of shoreline does California have?

Think and Grow: Modeling Real Life

Example The attendance on the second day of a music festival is 10,013 fewer people than on the third day. How many total people attend the three-day music festival?

Day	Attendance
1	76,914
2	85,212
3	?

Think: What do you know? What do you need to find? How will you solve?

Step 1: How many people attend the festival on the third day?

85,212	10,013

a

a is the unknown sum.

$85{,}212 + 10{,}013 = a$

$$\begin{array}{r} 85{,}212 \\ +\ 10{,}013 \\ \hline \square \end{array}$$

$a =$ ______

Remember, you can estimate $85{,}000 + 10{,}000 = 95{,}000$ to check whether your answer is reasonable.

Step 2: Use a to find the total attendance for the three-day festival.

$76{,}914 + 85{,}212 + a = f$

f is the unknown sum.

$76{,}914 + 85{,}212 +$ ______ $= f$

$$\begin{array}{r} 76{,}914 \\ 85{,}212 \\ +\ \square \\ \hline \square \end{array}$$

$f =$ ______

You can estimate $80{,}000 + 90{,}000 + 100{,}000 = 270{,}000$ to check whether your answer is reasonable.

______ total people attend the music festival.

Show and Grow I can think deeper!

8. A construction company uses 3,239 more bricks to construct Building 1 than Building 2. How many bricks does the company use to construct all three buildings?

Building	Bricks Used
1	11,415
2	?
3	16,352

Name ___________________________

Learning Target: Use the problem-solving plan to solve two-step addition and subtraction word problems.

Example A car owner wants to sell his car when the odometer reads 100,000 miles. The owner bought the car with 44,901 miles on it. He drives a total of 27,298 miles over the next two years. How many more miles can he drive before he sells his car?

Think: What do you know? What do you need to find? How will you solve?

Step 1:

44,901	27,298

k

k is the unknown sum.

$44{,}901 + 27{,}298 = k$

$$\begin{array}{r} 44{,}901 \\ +\ 27{,}298 \\ \hline \boxed{72{,}199} \end{array}$$

$k =$ 72,199

Step 2:

$k =$ 72,199	j

100,000

j is the unknown difference.

$100{,}000 -$ 72,199 $= j$

$$\begin{array}{r} 100{,}000 \\ -\ \boxed{72{,}199} \\ \hline \boxed{27{,}801} \end{array}$$

$j =$ 27,801

He can drive 27,801 more miles before he sells his car.

Understand the problem. Then make a plan. How will you solve? Explain.

1. A cargo plane weighs 400,000 pounds. After a load of cargo is removed, the plane weighs 336,985 pounds. Then a 12,395-pound load is removed. How many pounds of cargo are removed in all?

2. A ski resort uses 5,200 gallons of water per minute to make snow. A family uses 361 gallons of water each day. How many more gallons of water does the ski resort use to make snow in 2 minutes than a family uses in 1 day?

3. In July, a website receives 379,162 fewer orders than in May and June combined. The website receives 542,369 orders in May and 453,708 orders in June. How many orders does the website receive in July?

4. **Writing** Write and solve a two-step word problem that can be solved using addition or subtraction.

5. **Modeling Real Life** World War I lasted from 1914 to 1918. World War II lasted from 1939 to 1945. How much longer did World War II last than World War I?

6. **Modeling Real Life** Twenty people each donate $9 to a charity. Sixty people each donate $8. The charity organizer wants to raise a total of $1,500. How much more money does the organizer need to raise?

7. **DIG DEEPER!** The blackpoll warbler migrates 2,376 miles, stops, and then flies another 3,289 miles to reach its destination. The arctic tern migrates 11,013 miles, stops, and then flies another 10,997 miles to reach its destination. How much farther is the arctic tern's migration than the blackpoll warbler's migration?

Blackpoll Warbler

Artic Tern

Review & Refresh

Write the time. Write another way to say the time.

8.

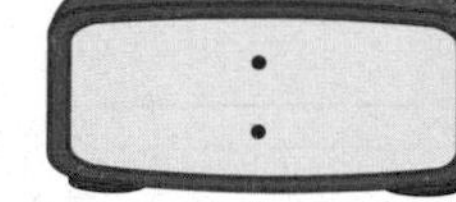

9.

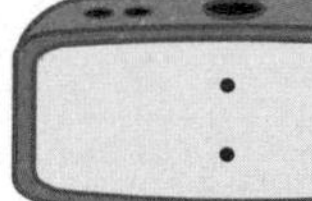

10.

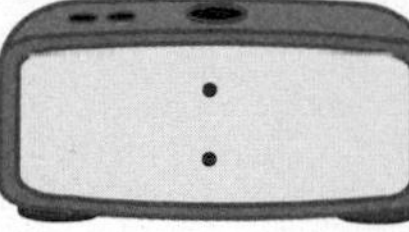

Name ______________________________

Performance Task 2

1. The time line shows the population of Austin, Texas from 1995 to 2015.

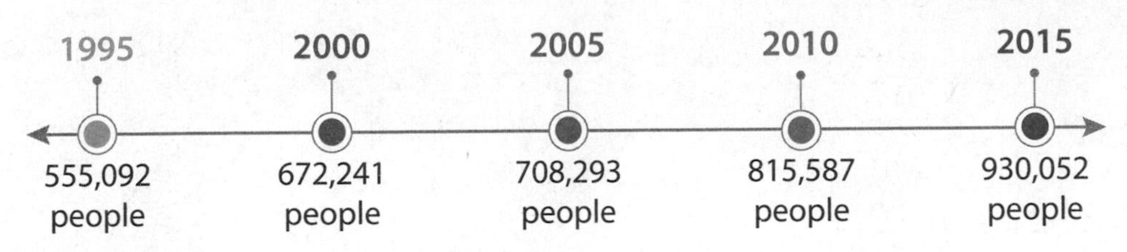

a. Which place would you round to when estimating population? Explain.

b. Estimate Austin's population each year on the timeline.

c. Use your estimates to complete the bar graph.

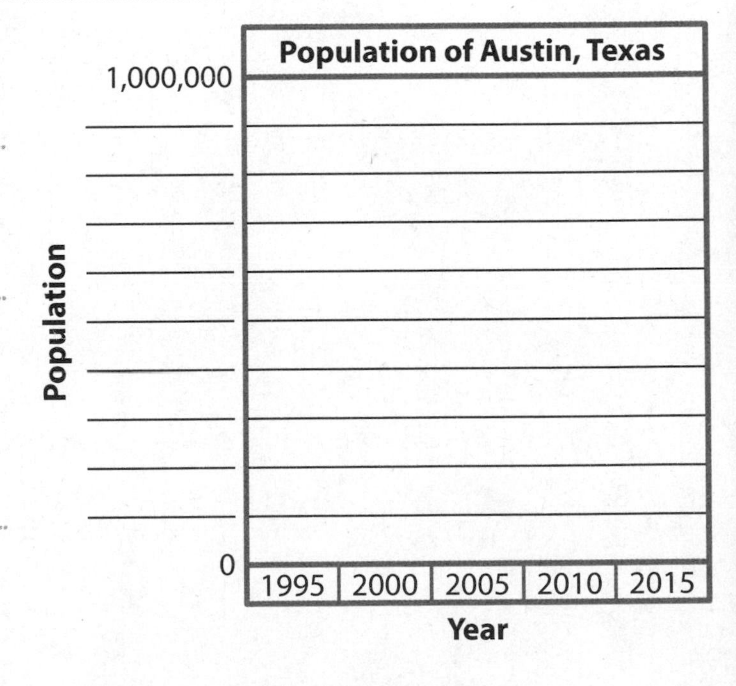

d. Between the years 1995 and 2015, did Austin's population increase, decrease, or stay the same? Explain.

e. During which five-year period did the population increase the most? Explain.

f. About how many more people lived in Austin in the year 2015 than in the year 1995?

g. Do you think the population will be more than 1,000,000 in the year 2020? Explain.

Race to the Moon

Directions:

1. Players take turns.
2. On your turn, flip a Race for the Moon Card and find the sum or difference.
3. Move your piece to the next number on the board that is highlighted in your answer.

3 Multiply by One-Digit Numbers

Chapter Learning Target:
Understand multiplying one-digit numbers.

Chapter Success Criteria:
- I can find the product of two numbers.
- I can use rounding to estimate a product.
- I can write multiplication problems.
- I can solve a problem using an equation.

- Have you ever seen a fireworks display? What types of events have fireworks displays?
- At a fireworks display, you see the lights before you hear the sounds because light travels faster than sound. How can you use multiplication to find out how far away you are from the fireworks?

Name ______________________________

3 Vocabulary

Review Words

Commutative Property of Multiplication
factors
product

Organize It

Use the review words to complete the graphic organizer.

[ ]

Changing the order of [ ] does not change the [ ].

$4 \times 3 = 12$ $3 \times 4 = 12$

So, $4 \times 3 = 3 \times 4$.

Define It

Use your vocabulary cards to match.

1. Distributive Property

2. partial products

```
   39
 × 7
   63   7 × 9
+ 210   7 × 30
  273
```

$3 \times (5 + 2) = (3 \times 5) + (3 \times 2)$

$3 \times (5 - 2) = (3 \times 5) - (3 \times 2)$

Chapter 3 Vocabulary Cards

Distributive Property

partial products

The products found by breaking apart a factor into ones, tens, hundreds, and so on, and multiplying each of these by the other factor

$$\begin{array}{r} 39 \\ \times\ 7 \\ \hline \boxed{63} \\ +\ \boxed{210} \\ \hline 273 \end{array} \quad \begin{array}{l} \\ \\ 7 \times 9 \\ 7 \times 30 \\ \\ \end{array}$$

partial products → 63, 210

$3 \times (5 + 2) = (3 \times 5) + (3 \times 2)$

$3 \times (5 - 2) = (3 \times 5) - (3 \times 2)$

Name ______________________________

Understand Multiplicative Comparisons

3.1

Learning Target: Use multiplication to compare two numbers.

Success Criteria:

- I can write addition or multiplication equations given a comparison sentence.
- I can write a comparison sentence given an addition or a multiplication equation.
- I can solve comparison word problems involving multiplication.

Model the counters. Draw to show your model.

There are 20 counters. Five of the counters are yellow. The rest are red.

 :

 :

How many more red counters are there than yellow counters?

How many times as many red counters are there as yellow counters?

Structure Explain how you can use an addition equation or a multiplication equation to compare the numbers of yellow counters and red counters.

Think and Grow: Understand Multiplicative Comparisons

You can use multiplication to compare two numbers.

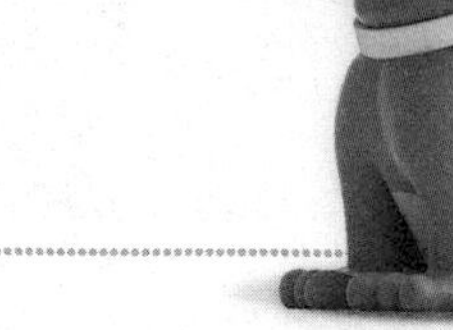

Remember, you can use the Commutative Property of Multiplication to multiply in any order.

Example Write two comparison sentences for $24 = 4 \times 6$.

6			
6	6	6	6

24

24 is ______ times as many as ______.

4					
4	4	4	4	4	4

24

24 is ______ times as many as ______.

You can compare two numbers using addition or multiplication.

- Use addition to find *how many more* or *how many fewer*.
- Use multiplication to find *how many times* as much.

Example Write an equation for each comparison sentence.

12 is 8 more than 4.

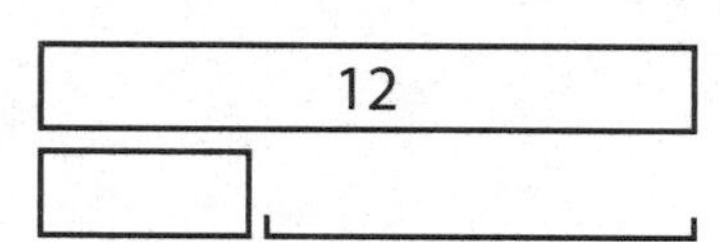

$12 =$ ______ $+$ ______

12 is 3 times as many as 4.

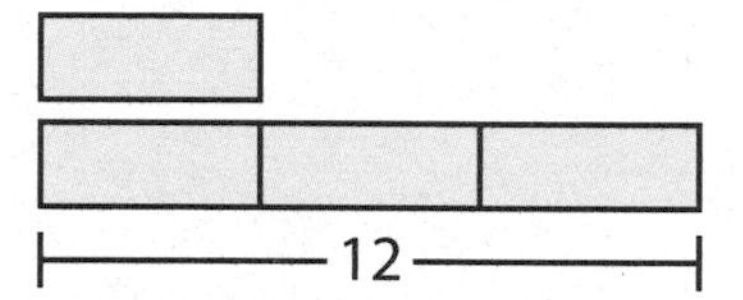

$12 =$ ______ $\times$ ______

Show and Grow I can do it!

Write two comparison sentences for the equation.

1. $15 = 3 \times 5$

2. $32 = 4 \times 8$

Draw a model for the comparison sentence. Then write an equation.

3. 21 is 14 more than 7.

4. 40 is 8 times as many as 5.

Name ______________________________

Apply and Grow: Practice

Write two comparison sentences for the equation.

5. $48 = 6 \times 8$

6. $63 = 7 \times 9$

Write an equation for the comparison sentence.

7. 20 is 2 times as many as 10.

8. 18 is 10 more than 8.

9. 35 is 7 times as many as 5.

10. 16 is 4 times as many as 4.

11. Earthworms have four more hearts than humans. How many hearts do earthworms have?

12. Ants can lift 50 times their body weight. An ant weighs 5 milligrams. How much weight can the ant lift?

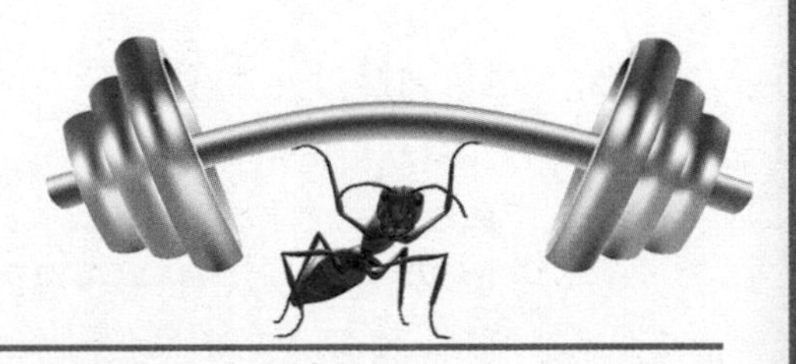

13. **Writing** Explain how you know the statement "32 is 8 times as many as 4" is a comparison involving multiplication.

14. MP **Number Sense** Write an addition comparison statement and a multiplication comparison statement for the numbers 8 and 24.

Think and Grow: Modeling Real Life

Example You perform a science experiment and use 4 times as much hydrogen peroxide as water. You use a total of 10 tablespoons of liquid. How many tablespoons of hydrogen peroxide do you use?

Draw a model.

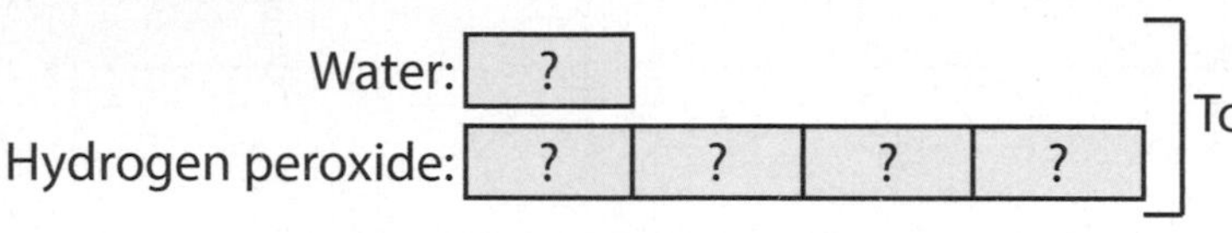

Total = 10 tablespoons of liquid

Find the number of tablespoons of water.

The model shows ______ equal parts. There are ______ tablespoons of liquid in all.

$5 \times ? = 10$ Think: 5 times what number equals 10?

You use ______ tablespoons of water.

Find the number of tablespoons of hydrogen peroxide.

You use ______ times as much hydrogen peroxide as water.

$2 \times 4 =$ ______

So, you use ______ tablespoons of hydrogen peroxide.

Show and Grow *I can think deeper!*

15. A bicycle-sharing station on Main Street has 5 times as many bicycles as a station on Park Avenue. There are 24 bicycles at the two stations. How many bicycles are at the Main Street station?

16. In the 2016 Olympics, Brazil won 6 silver medals. France won 3 times as many silver medals as Brazil. How many silver medals did France win?

17. Of all the national flags in the world, there are 3 times as many red, white, and blue flags as there are red, white, and green flags. There are 40 flags with these color combinations. How many more flags are red, white, and blue than red, white, and green?

Name ______________________

Homework & Practice 3.1

Learning Target: Use multiplication to compare two numbers.

Example Write two comparison sentences for $6 = 2 \times 3$.

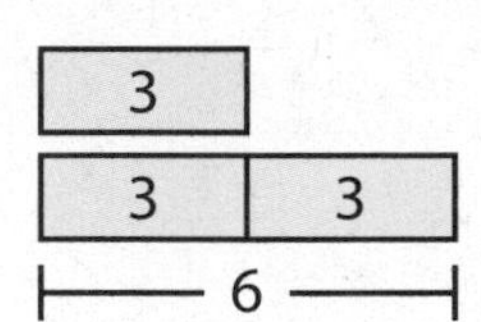

2
2 2 2
6

6 is __2__ times as many as __3__.

6 is __3__ times as many as __2__.

Example Write an equation for each comparison sentence.

18 is 12 more than 6.

18
6 12

$18 =$ __6__ $+$ __12__

18 is 3 times as many as 6.

6
6 6 6
18

$18 =$ __3__ $\times$ __6__

Write two comparison sentences for the equation.

1. $24 = 8 \times 3$

2. $14 = 7 \times 2$

Write an equation for the comparison sentence.

3. 30 is 6 times as many as 5.

4. 27 is 3 times as many as 9.

5. 12 is 7 more than 5.

6. 10 is 2 times as many as 5.

7. The House of Representatives has 335 more members than the Senate. The Senate has 100 members. How many members does the House of Representatives have?

8. A lion's roar can be heard 5 miles away. The vibrations from an elephant's stomp can be felt 4 times as many miles away as the lion's roar can be heard. How many miles away can the vibrations be felt?

9. MP **Reasoning** Newton says the equation $270 = 30 \times 9$ means 270 is 30 times as many as 9. Descartes says it means 270 is 9 times as many as 30. Explain how you know they are both correct.

10. MP **Precision** Compare the door's height to the desk's height using multiplication and addition.

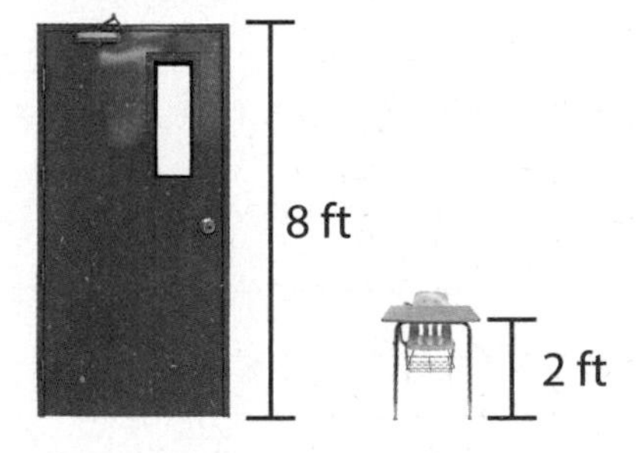

11. **Open-Ended** Write a comparison statement for a sum of 28.

12. **Modeling Real Life** There are 12 shepherds and retrievers in all at a dog park. There are 2 times as many shepherds as retrievers. How many retrievers are there?

13. **Modeling Real Life** Pythons sleep 6 times as long as horses. Horses sleep 3 hours each day. How many hours do pythons sleep each day?

14. DIG DEEPER! You have 8 times as many dimes as nickels. You have 18 dimes and nickels altogether. How much money do you have in all?

Review & Refresh

Find the missing factor.

15. $7 \times \_\_\_\_ = 280$

16. $\_\_\_\_ \times 30 = 270$

17. $8 \times \_\_\_\_ = 640$

18. $\_\_\_\_ \times 90 = 540$

19. $2 \times \_\_\_\_ = 40$

20. $\_\_\_\_ \times 50 = 350$

Name ______________________________

Multiply Tens, Hundreds, and Thousands

3.2

Learning Target: Use place value to multiply by tens, hundreds, or thousands.

Success Criteria:

- I can find the product of a one-digit number and a multiple of ten, one hundred, or one thousand.
- I can describe a pattern when multiplying by tens, hundreds, or thousands.

Explore and Grow

Use models to find each product. Draw your models.

$4 \times 3 =$ ____	$4 \times 30 =$ ____
$4 \times 300 =$ ____	$4 \times 3{,}000 =$ ____

What pattern do you notice?

Repeated Reasoning How does 3×7 help you to find $3 \times 7{,}000$? Explain.

Think and Grow: Multiply Tens, Hundreds, and Thousands

You can use place value to multiply by tens, hundreds, or thousands.

Example Find each product.

$7 \times 200 = 7 \times$ 2 hundreds

$=$ ______ hundreds

$=$ ______

So, $7 \times 200 =$ ______.

$3 \times 4{,}000 = 3 \times$ ______ thousands

$=$ ______ thousands

$=$ ______

So, $3 \times 4{,}000 =$ ______.

Example Find each product.

$8 \times 5 = 40$	Multiplication fact
$8 \times 50 = 400$	Find 8×5; write 1 zero to show tens.
$8 \times 500 =$ 4,000	Find 8×5; write 2 zeros to show hundreds.
$8 \times 5{,}000 =$ 40,000	Find 8×5; write 3 zeros to show thousands.

Notice the Pattern: Write the multiplication fact and the same number of zeros that are in the second factor.

Show and Grow *I can do it!*

Find each product.

1. $6 \times 9 =$ 54

$6 \times 90 =$ 540

$6 \times 900 =$ 5,400

$6 \times 9{,}000 =$ 54,000

2. $5 \times 2 =$ 10

$5 \times 20 =$ 100

$5 \times 200 =$ 1,000

$5 \times 2{,}000 =$ 10,000

3. $2 \times 3 =$ 6

$2 \times 300 =$ 600

4. $9 \times 8 =$ 72

$9 \times 8{,}000 =$ 72,000

Name ______________________________

Apply and Grow: Practice

Find the product.

5. 7 × 700 = ______	**6.** 3 × 100 = ______	**7.** 8,000 × 3 = ______
8. 60 × 2 = ______	**9.** 4 × 4,000 = ______	**10.** 700 × 5 = ______
11. 900 × 7 = ______	**12.** 50 × 3 = ______	**13.** 1,000 × 8 = ______

Find the missing factor.

14. ______ × 6,000 = 24,000	**15.** 9 × ______ = 450	**16.** ______ × 30 = 210
17. 2 × ______ = 800	**18.** ______ × 1,000 = 9,000	**19.** 8 × ______ = 640

Compare.

20. 7 × 60 ◯ 400	**21.** 500 × 4 ◯ 2,000	**22.** 3 × 9,000 ◯ 39,000

23. The North Canadian River is 800 miles long. The Amazon River is 5 times longer than the North Canadian River. How many miles long is the Amazon River?

24. One reusable bag can prevent the use of 600 plastic bags. Six reusable bags can prevent the use of how many plastic bags?

25. **YOU BE THE TEACHER** Your friend says the product of 6 and 500 will have 2 zeros. Is your friend correct? Explain.

Think and Grow: Modeling Real Life

Example An aquarium has 7 bottlenose dolphins. Each dolphin eats 60 pounds of fish each day. The aquarium has 510 pounds of fish. Does the aquarium have enough fish to feed the dolphins?

Think: What do you know? What do you need to find? How will you solve?

Step 1: How many pounds of fish do all of the dolphins eat?

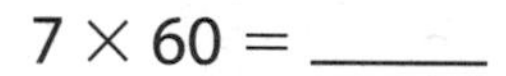

$7 \times 60 =$ ____

All of the dolphins eat ____ pounds of fish.

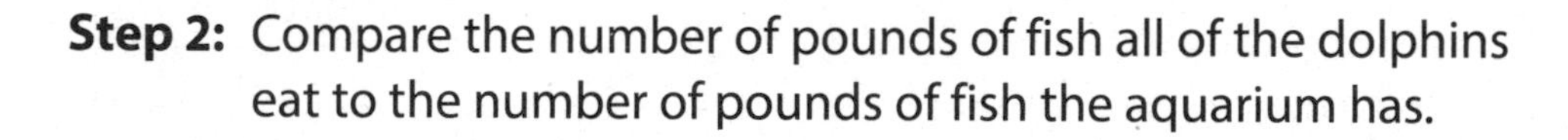

Step 2: Compare the number of pounds of fish all of the dolphins eat to the number of pounds of fish the aquarium has.

The aquarium ____________ have enough fish to feed the dolphins.

Show and Grow *I can think deeper!*

26. Students want to make 400 dream catchers for a craft fair. Each dream catcher needs 8 feathers. The students have 3,100 feathers. Do the students have enough feathers for all of the dream catchers?

27. A principal has 3 rolls of 800 raffle tickets each and 5 rolls of 9,000 raffle tickets each. How many raffle tickets does the principal have?

28. You have 2 sheets of 4 stickers each. Your friend has 20 times as many stickers as you. Your teacher has 700 times as many stickers as you. How many stickers do the three of you have in all?

Name ________________________________

Estimate Products by Rounding 3.3

Learning Target: Use rounding to estimate products.

Success Criteria:

- I can use rounding to estimate a product.
- I can find two estimates that a product is between.
- I can tell whether a product is reasonable.

There are 92 marbles in each jar.

Estimate the total number of marbles in two ways.

Round 92 to the nearest ten.

Round 92 to the nearest hundred.

Which estimate do you think is closer to the total number of marbles? Explain.

Repeated Reasoning Would your answer to the question above change if there were 192 marbles in each jar? Explain.

Think and Grow: Estimate Products

You can estimate a product by rounding.

Remember, an estimate is a number that is close to an exact number.

Example Estimate 7×491.

Round 491 to the nearest hundred. Then multiply.

7×491

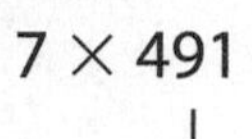

$7 \times$ ______ $=$ ______

So, 7×491 is about ______.

When solving multiplication problems, you can check whether an answer is reasonable by finding two estimates that a product is between.

Example Find two estimates that the product of 4×76 is between.

Think: 76 is between 70 and 80.

4×76 → $4 \times 70 =$ ______

4×76 → $4 \times 80 =$ ______

So, the product is between ______ and ______.

Show and Grow *I can do it!*

Estimate the product.

1. 3×89

2. 8×721

3. $5 \times 7{,}938$

Find two estimates that the product is between.

4. 9×44

5. 2×657

6. $6 \times 4{,}243$

Name ________________________________

Apply and Grow: Practice

Estimate the product.

7. 4×65

8. 248×7

9. $3 \times 9{,}032$

Find two estimates that the product is between.

10. 32×9

11. 970×5

12. $6 \times 5{,}328$

13. A sales representative sells 4 smart watches for $199 each. To determine whether the sales representative collects at least $1,000, can you use an estimate, or is an exact number required? Explain.

YOU BE THE TEACHER A student finds the product. Is her answer reasonable? Estimate to check.

14. $480 \times 8 \stackrel{?}{=} 43{,}840$

15. $1{,}904 \times 4 \stackrel{?}{=} 7{,}616$

MP Number Sense Use two of the numbers below to write an expression for a product that can be estimated as shown. You may use the numbers more than once.

5 8,932 8 4,592 3 2,842

16. Estimate: 40,000

_______ × _______

17. Estimate: 9,000

_______ × _______

18. Estimate: 72,000

_______ × _______

19. Estimate: 45,000

_______ × _______

Think and Grow: Modeling Real Life

Example About how much more money was earned from new release rentals than from top pick rentals?

Online Video Rentals

Type of video	Number rented
New Release $5	3,047
Top Pick $3	3,986
Television Episode $2	849

Number rented (0 to 5,000); Type of video

Estimate the amount of money earned from new release rentals.

$\$5 \times 3{,}047$

↓

$5 × ______ = $______

Estimate the amount of money earned from top pick rentals.

$\$3 \times 3{,}986$

↓

$3 × ______ = $______

Subtract.

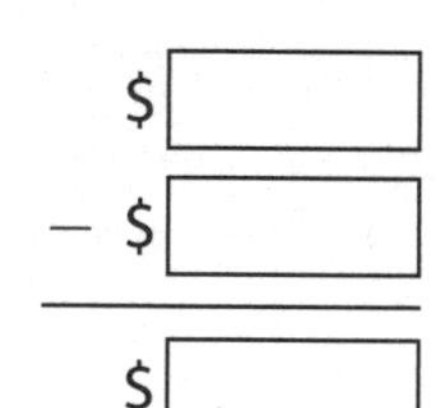

$ ☐
− $ ☐
$ ☐

So, about $______ more was earned from new release rentals.

Show and Grow *I can think deeper!*

Use the graph above.

20. About how much more money was earned from top pick rentals than from television episode rentals?

21. Your friend says that the number of top picks rented is about 5 times as many as the number of television episodes rented. Is your friend correct? Explain.

22. An accountant says that the total amount of money earned from new release rentals is $11,958. Check whether the accountant's answer is reasonable by finding two estimates that the total is between.

Name ______________________________

Learning Target: Use rounding to estimate products.

Example Estimate $5 \times 3{,}285$.

Round 3,285 to the nearest thousand. Then multiply.

$5 \times 3{,}285$

↓

$5 \times \underline{3{,}000} = \underline{15{,}000}$

So, $5 \times 3{,}285$ is about $\underline{15{,}000}$.

Example Estimate 8×671 by finding two numbers the product is between.

Think: 671 is between 600 and 700.

8×671 → $8 \times 600 = \underline{4{,}800}$

8×671 → $8 \times 700 = \underline{5{,}600}$

So, the product is between $\underline{4{,}800}$ and $\underline{5{,}600}$.

Estimate the product.

1. 7×42

2. 85×4

3. 2×698

4. 6×705

5. $1{,}834 \times 9$

6. $7{,}923 \times 8$

Find two estimates that the product is between.

7. 3×95

8. 23×5

9. 537×6

10. 8×309

11. $1{,}649 \times 7$

12. $4 \times 6{,}203$

13. Construction workers build 54 feet of a bridge each day for 9 days. To determine whether the bridge is at least 490 feet long, can you use an estimate, or is an exact answer required? Explain.

14. **YOU BE THE TEACHER** A student finds the product. Is his answer reasonable? Estimate to check.

$$5{,}692 \times 3 \stackrel{?}{=} 17{,}076$$

15. **DIG DEEPER!** An astronaut earns $5,428 each month. You estimate that she will earn $35,000 in 7 months. Is the amount she earns in 7 months greater than or less than your estimate? Explain.

16. **Writing** Explain how estimating by rounding can be helpful as a check when finding a product.

Modeling Real Life Use the graph to answer the question.

17. Which animal has a heart rate that is about 2 times as fast as a whale's?

18. About how many times does a giraffe's heart beat in 5 minutes? Find two estimates that the answer is between.

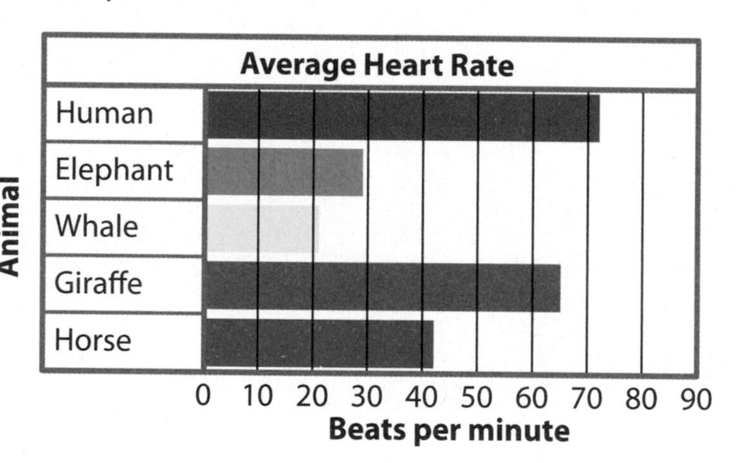

Review & Refresh

Use the Distributive Property to find the product.

19. $9 \times 6 = \_\_\_ \times (5 + \_\_\_)$

$= (9 \times 5) + (\_\_\_ \times \_\_\_)$

$= \_\_\_ + \_\_\_$

$= \_\_\_$

20. $7 \times 7 = 7 \times (5 + \_\_\_)$

$= (\_\_\_ \times 5) + (7 \times \_\_\_)$

$= \_\_\_ + \_\_\_$

$= \_\_\_$

Name ______________________________

Use the Distributive Property to Multiply 3.4

Learning Target: Use the Distributive Property to multiply.

Success Criteria:

- I can draw an area model to multiply.
- I can use known facts to find a product.
- I can explain how to use the Distributive Property.

Explore and Grow

Use base ten blocks to model 4 × 16. Draw your model. Then find the area of the model.

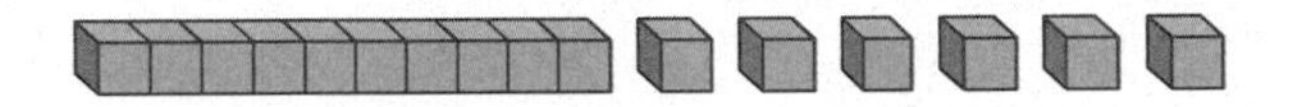

4 × 16 = _____

Break apart 16 to show two smaller models. Find the area of each model. What do you notice about the sum of the areas?

Area = _____ Area = _____

Reasoning How does this strategy relate to the Distributive Property? Explain.

Think and Grow: Use the Distributive Property to Multiply

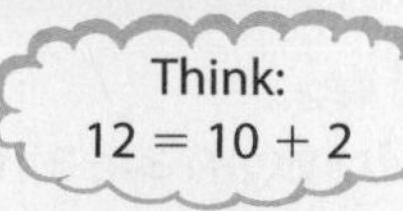

One way to multiply a two-digit number is to first break apart the number. Then use the Distributive Property.

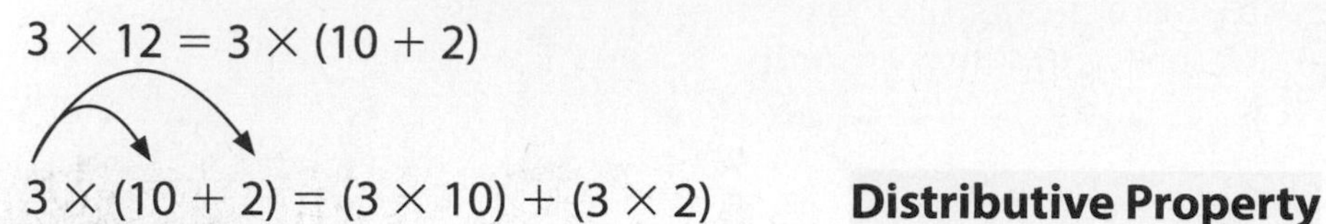

Example Use an area model to find 6×18.

Model the expression. Break apart 18 as 10 + 8.

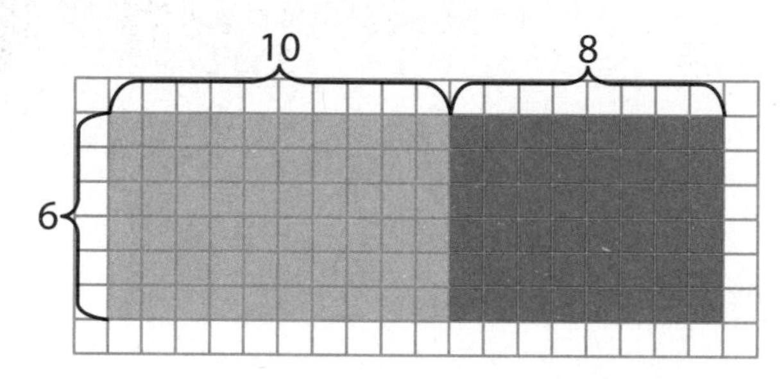

$6 \times 18 = 6 \times (\_\_\_\_ + \_\_\_\_)$ Rewrite 18 as 10 + 8.

$= (6 \times 10) + (6 \times 8)$ Distributive Property

$= 60 + 48$

$= \_\_\_\_$

Show and Grow *I can do it!*

Draw an area model. Then find the product.

1. $7 \times 11 =$ 77

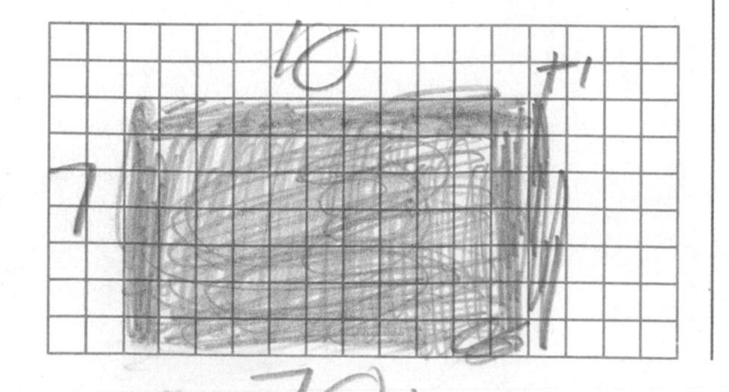

2. $2 \times 15 =$ 30

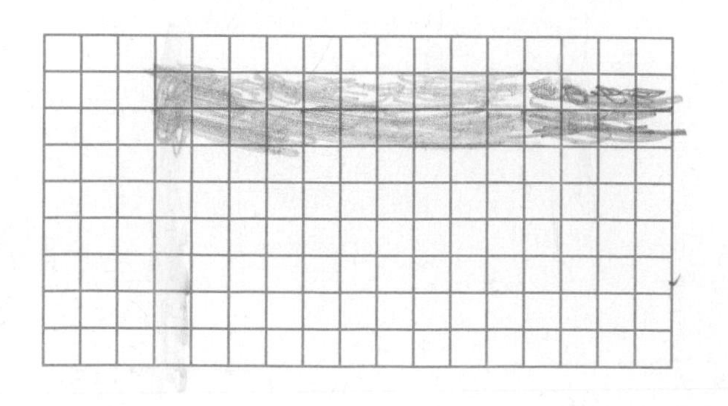

Name ______________________________

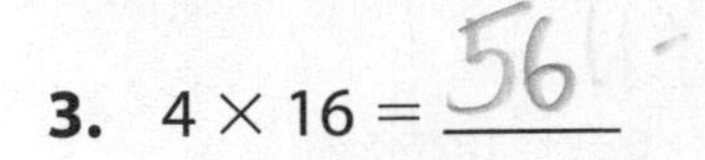

Apply and Grow: Practice

Draw an area model. Then find the product.

3. $4 \times 16 =$ ______

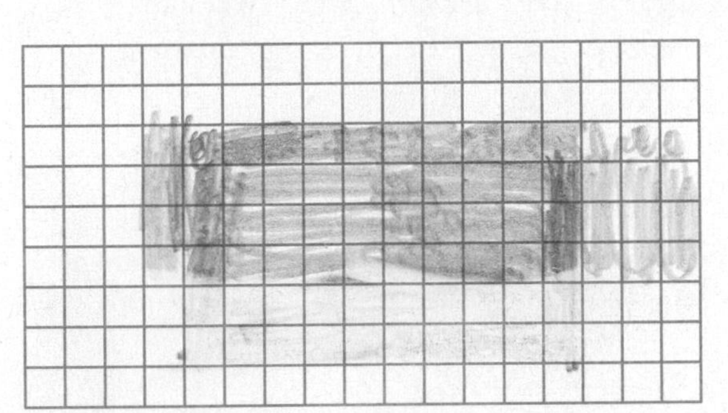

4. $9 \times 18 =$ ______

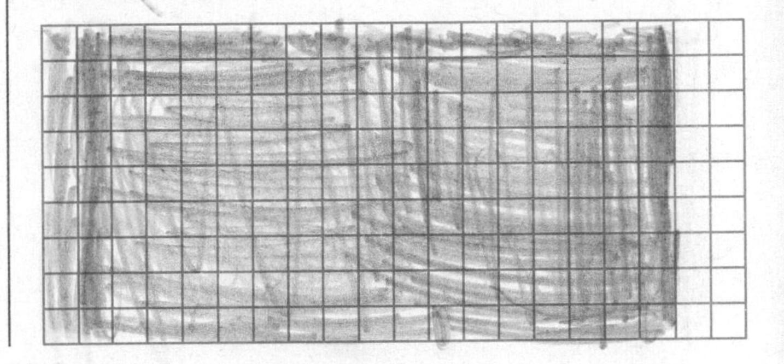

5. $6 \times 27 =$ ______

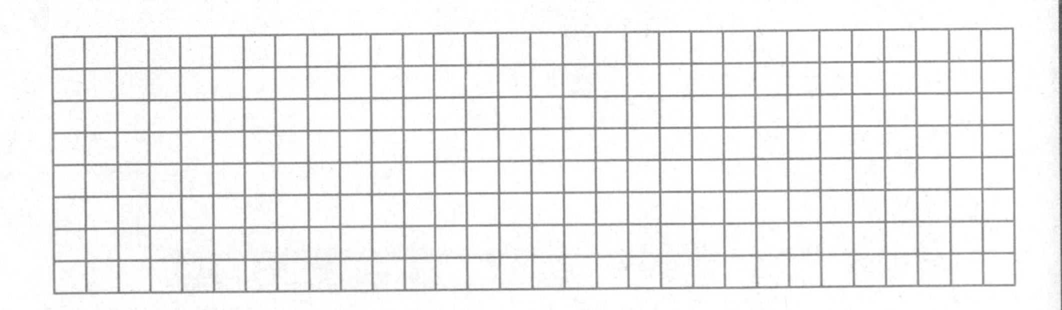

Find the product.

6. $3 \times 46 =$ ______

7. $8 \times 35 =$ ______

8. $5 \times 72 =$ ______

9. MP **Number Sense** Use the area model to complete the equation.

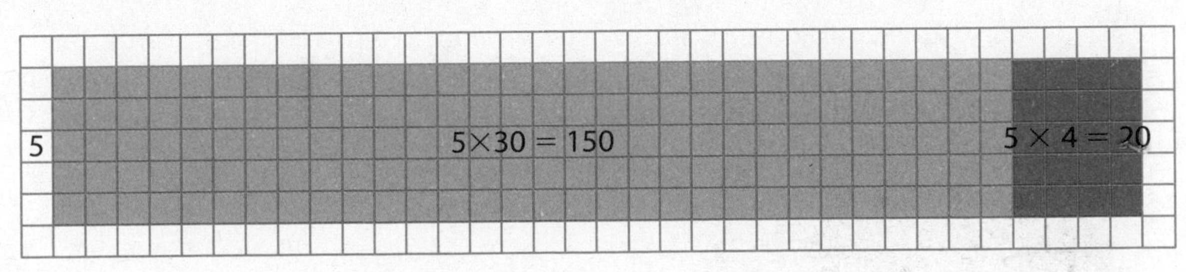

$5 \times$ ______ $=$ ______

Think and Grow: Modeling Real Life

Example A parking garage has 9 floors. Each floor has 78 parking spaces. 705 cars are trying to park in the garage. Are there enough parking spaces? Explain.

Multiply the number of floors by the number of parking spaces on each floor.

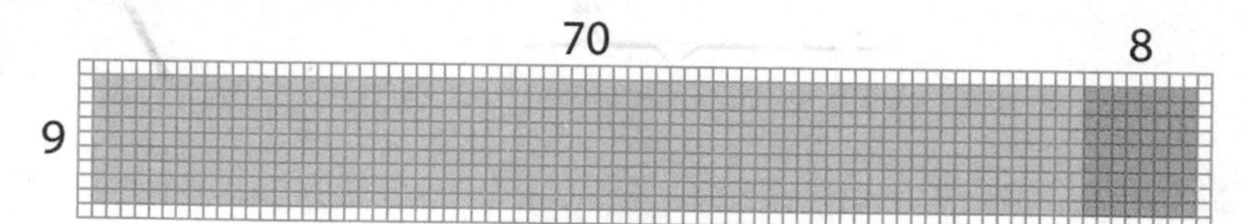

$9 \times 78 = 9 \times (70 + 8)$

$= (9 \times 70) + (9 \times 8)$

$=$ ______ $+$ ______

$=$ ______

Compare the number of spaces to the number of cars trying to park in the garage.

There __________ enough parking spaces.

Explain:

Show and Grow *I can think deeper!*

10. Your piano teacher wants you to practice playing the piano 160 minutes this month. So far, you have practiced 5 minutes each day for 24 days. Have you reached the goal your teacher set for you? Explain.

11. Newton has $150. He wants to buy a bicycle that costs 4 times as much as the helmet. Does he have enough money to buy the bicycle and the helmet? Explain.

$28

12. A baby orangutan weighs 3 pounds. The mother orangutan weighs 27 times as much as the baby. The father orangutan weighs 63 times as much as the baby. What are the weights of the mother and the father?

Name ______________________

Homework & Practice 3.4

Learning Target: Use the Distributive Property to multiply.

Example Use an area model to find 8×24.

Model the expression. Break apart 24 as $20 + 4$.

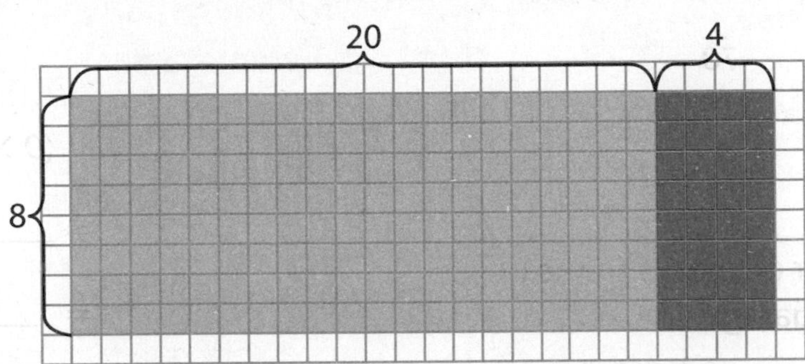

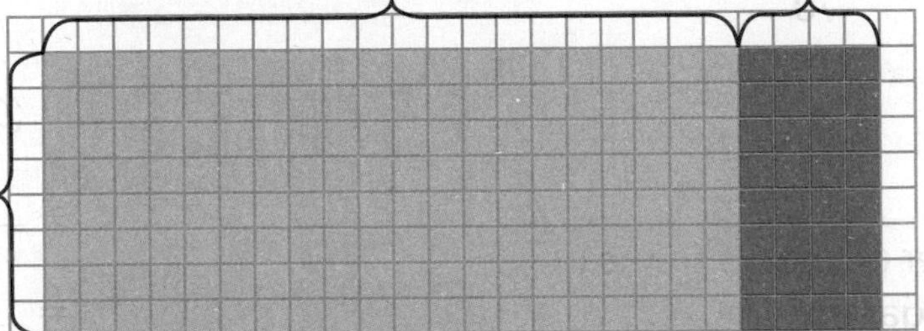

$8 \times 24 = 8 \times (\underline{\ 20\ } + \underline{\ 4\ })$ Rewrite 24 as $20 + 4$.

$= (8 \times 20) + (8 \times 4)$ Distributive Property

$= 160 + 32$

$= \underline{\ 192\ }$

Draw an area model. Then find the product.

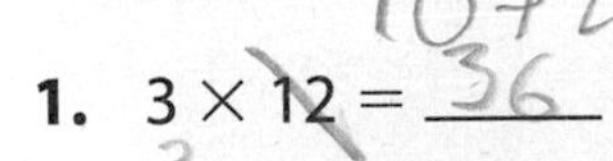
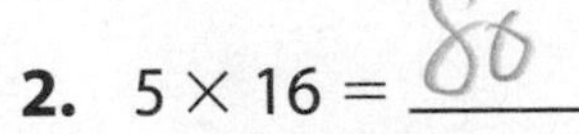

1. $3 \times 12 =$ ____

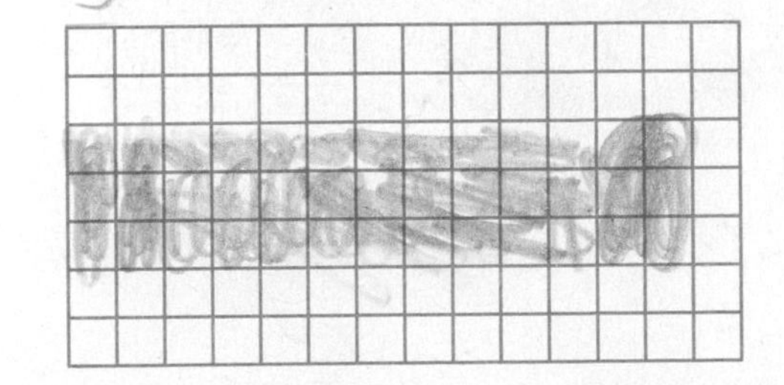

2. $5 \times 16 =$ ____

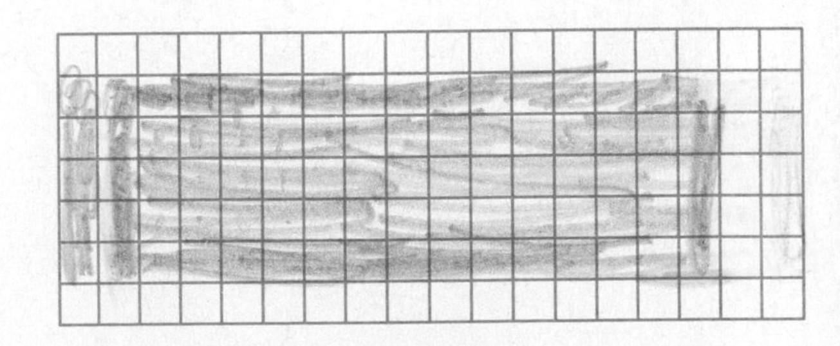

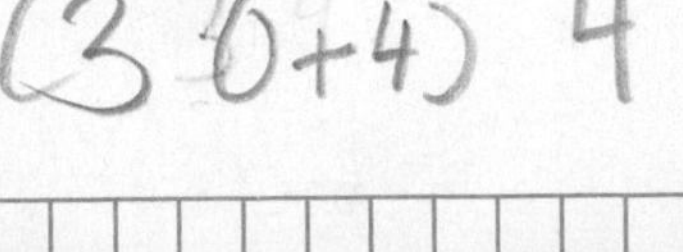
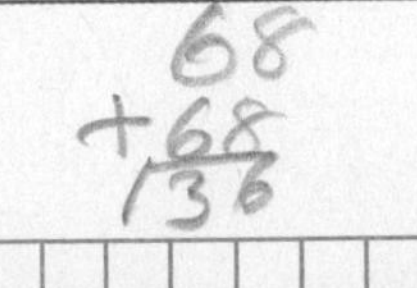

3. $4 \times 34 =$ ____

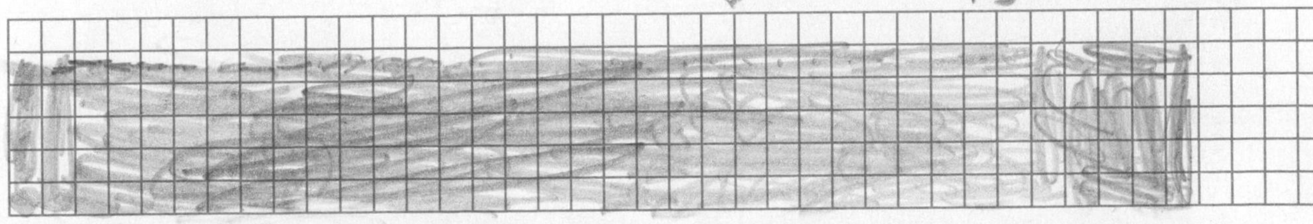

Find the product.

4. $9 \times 56 =$ ______

5. $71 \times 2 =$ ______

6. $3 \times 77 =$ ______

7. **Writing** Explain how you can use the Distributive Property to find a product.

8. **MP Structure** Use the Distributive Property to find 6×18 two different ways.

9. **MP Reasoning** To find 4×22, would you rather break apart the factor 22 as $20 + 2$ or as $11 + 11$? Explain.

10. **Modeling Real Life** The diameter of a firework burst, in feet, is 45 times the height of the shell in inches. You have an 8-inch firework shell. Will the shell produce a firework burst that has a diameter greater than 375 feet? Explain.

11. **Modeling Real Life** A juvenile bearded dragon should eat 48 crickets each day. You have 150 crickets. Do you have enough crickets to feed 3 juvenile bearded dragons? Explain.

Review & Refresh

Plot the fraction on a number line.

12. $\frac{7}{4}$

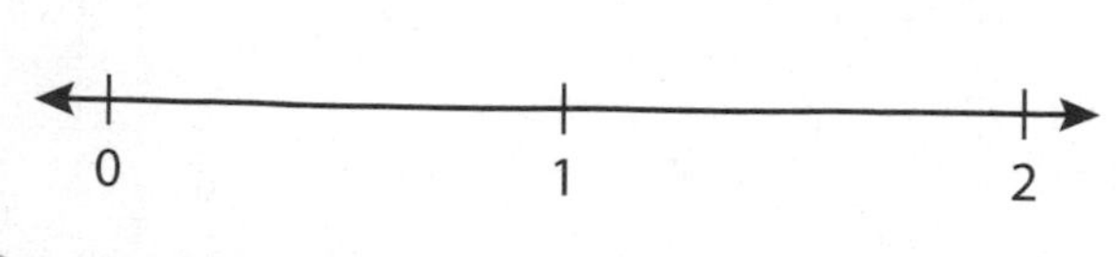

13. $\frac{4}{3}$

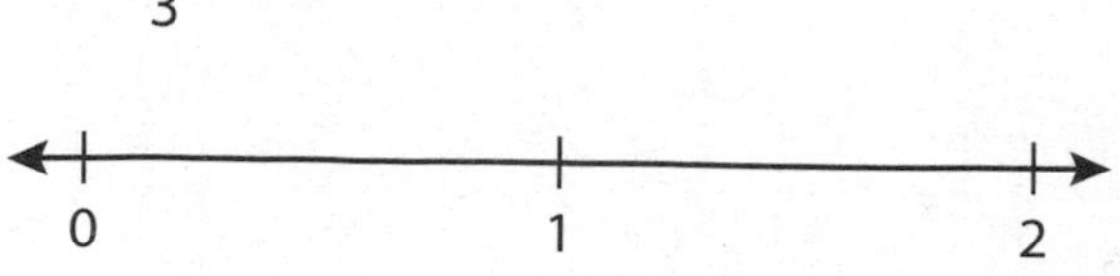

Name ________________________________

Use Expanded Form to Multiply 3.5

Learning Target: Use expanded form and the Distributive Property to multiply.

Success Criteria:

- I can use an area model to multiply.
- I can use expanded form and the Distributive Property to find a product.

Write 128 in expanded form.

$128 = \_\_\_\_\_ + \_\_\_\_\_ + \_\_\_\_\_$

Use expanded form to label the area model for 5×128. Find the area of each part.

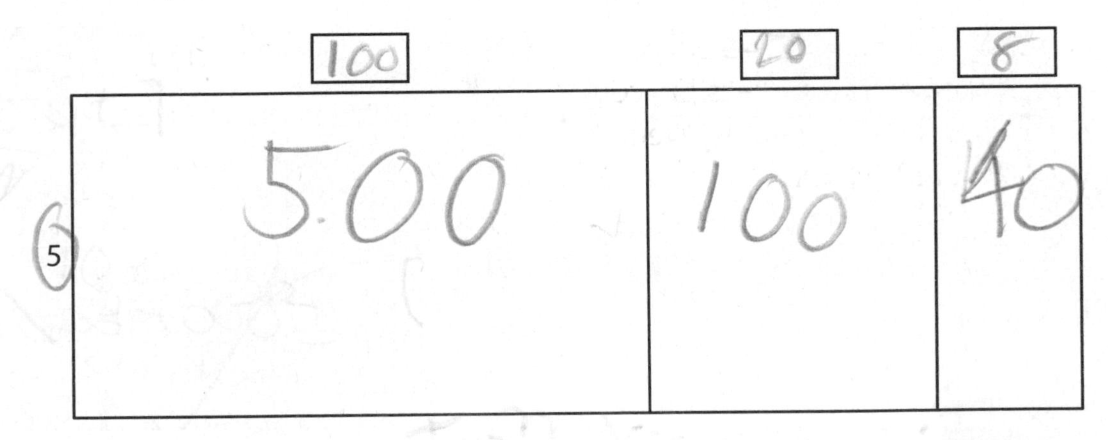

What is the sum of all of the parts? How does the sum relate to the product of 5 and 128?

Repeated Reasoning Explain to your partner how you can use expanded form to find 364×8.

Think and Grow: Use Expanded Form to Multiply

You can use expanded form and the Distributive Property to multiply.

Example Find 8×74.

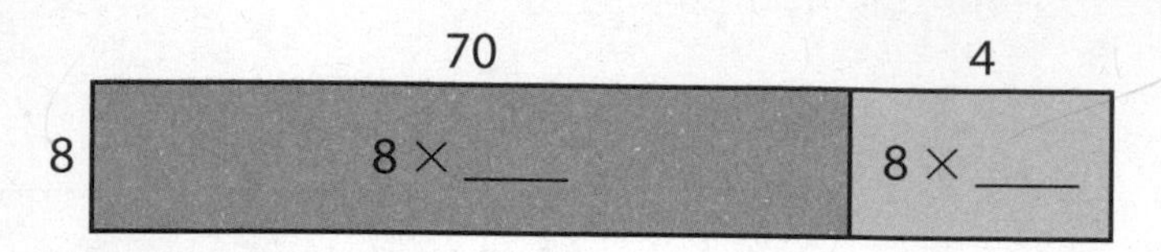

$8 \times 74 = 8 \times (70 + 4)$ Write 74 in expanded form.

$= (8 \times 70) + (8 \times 4)$ Distributive Property

$=$ ____ $+$ ____

$=$ ____

So, $8 \times 74 =$ ____.

Example Find $2 \times 5{,}607$.

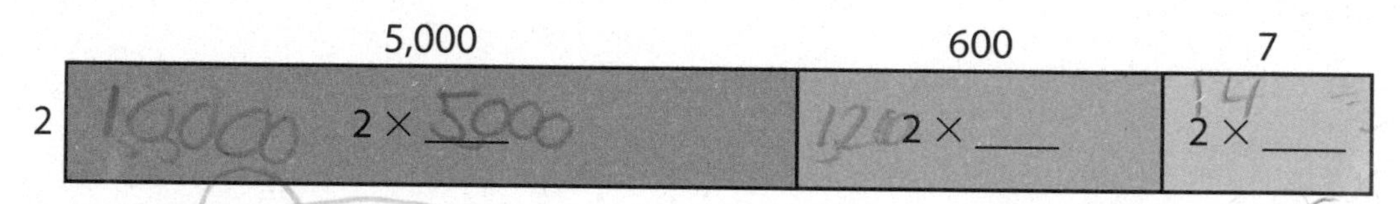

$2 \times 5{,}607 = 2 \times (5{,}000 + 600 + 7)$ Write 5,607 in expanded form.

$= (2 \times 5{,}000) + (2 \times 600) + (2 \times 7)$ Distributive Property

$=$ ____ $+$ ____ $+$ ____

$=$ ____

So, $2 \times 5{,}607 =$ ____.

Show and Grow **I can do it!**

Find the product.

1. $4 \times 306 = 4 \times ($ ____ $+$ ____ $)$

$= (4 \times$ ____ $) + (4 \times$ ____ $)$

$=$ ____ $+$ ____

$=$ ____

2. 7×549

Name ______________________________

Apply and Grow: Practice

Find the product.

3. $6 \times 85 =$ _____

	80	5
6	6 × _____	6 × _____

4. $2 \times 932 =$ _____

	900	30	2
2	2 × _____	2 × _____	2 × _____

5. $4 \times 690 = 4 \times (\_\_\_\_\_ + \_\_\_\_\_)$

$= (4 \times \_\_\_\_\_) + (4 \times \_\_\_\_\_)$

$= \_\_\_\_\_ + \_\_\_\_\_$

$= \_\_\_\_\_$

6. $1{,}027 \times 9 = (\_\_\_\_\_ + \_\_\_\_\_ + \_\_\_\_\_) \times 9$

$= (\_\_\_\_\_ \times 9) + (\_\_\_\_\_ \times 9) + (\_\_\_\_\_ \times 9)$

$= \_\_\_\_\_ + \_\_\_\_\_ + \_\_\_\_\_$

$= \_\_\_\_\_$

7. $487 \times 5 =$ _____

8. $8 \times 2{,}483 =$ _____

9. A basketball player made 269 three-point shots in a season. How many points did he score from three-point shots?

10. Your cousin runs 6 miles each week. There are 5,280 feet in a mile. How many feet does your cousin run each week?

11. **YOU BE THE TEACHER** Your friend finds 744×3. Is your friend correct? Explain.

$744 \times 3 = (700 + 40 + 4) \times 3$

$= (700 \times 3) + (40 \times 3) + (4 \times 3)$

$= 2{,}100 + 120 + 12$

$= 2{,}232$

12. **DIG DEEPER!** What is the greatest possible product of a two-digit number and a one-digit number? Explain.

96

Think and Grow: Modeling Real Life

Example A baby hippo is fed 77 fluid ounces of milk 5 times each day. There are 128 fluid ounces in 1 gallon. Are 2 gallons of milk enough to feed the baby hippo for 1 day?

Find the number of fluid ounces of milk the baby hippo drinks in 1 day.

$5 \times 77 = 5 \times (70 + 7)$

$= (5 \times 70) + (5 \times 7)$

$=$ _____ $+$ _____

$=$ _____

Find the number of fluid ounces in 2 gallons.

$2 \times 128 = 2 \times (100 + 20 + 8)$

$= (2 \times 100) + (2 \times 20) + (2 \times 8)$

$=$ _____ $+$ _____ $+$ _____

$=$ _____

Compare the products.

So, 2 gallons of milk __________ enough to feed the baby hippo for 1 day.

Show and Grow *I can think deeper!*

13. A school with 6 grades goes on a field trip. There are 64 students in each grade. One bus holds 48 students. Will 8 buses hold all of the students?

14. The average life span of a firefly is 61 days. The average life span of a Monarch butterfly is 4 times as long as that of a firefly. How many days longer is the average lifespan of a Monarch butterfly than a firefly?

15. A tourist is in Denver. His car can travel 337 miles using 1 tank of gasoline. He wants to travel to a city using no more than 3 tanks of gasoline. To which cities could the tourist travel?

City	Distance from Denver
Los Angeles	1,016 miles
Phoenix	821 miles
Dallas	796 miles

Name ____________________

Homework & Practice 3.5

Learning Target: Use expanded form and the Distributive Property to multiply.

Example Find 5×391.

	300	90	1
5	$5 \times$ 300	$5 \times$ 90	$5 \times$ 1

$5 \times 391 = 5 \times (300 + 90 + 1)$ — Write 391 in expanded form.

$= (5 \times 300) + (5 \times 90) + (5 \times 1)$ — Distributive Property

$= 1{,}500 + 450 + 5$

$= 1{,}955$

So, $5 \times 391 = 1{,}955$.

Find the product.

1. $7 \times 803 =$ ____

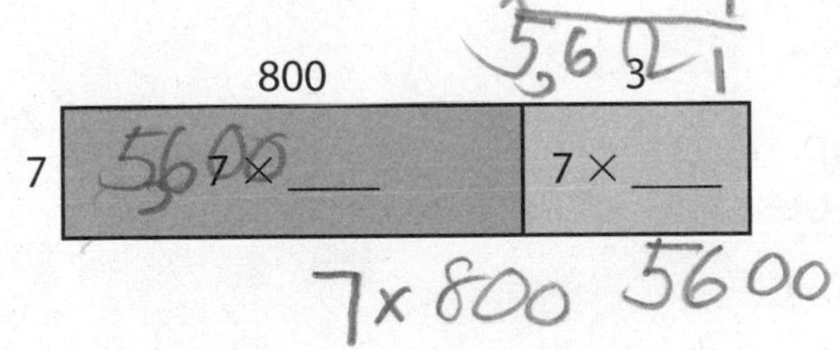

2. $9 \times 1{,}024 =$ ____

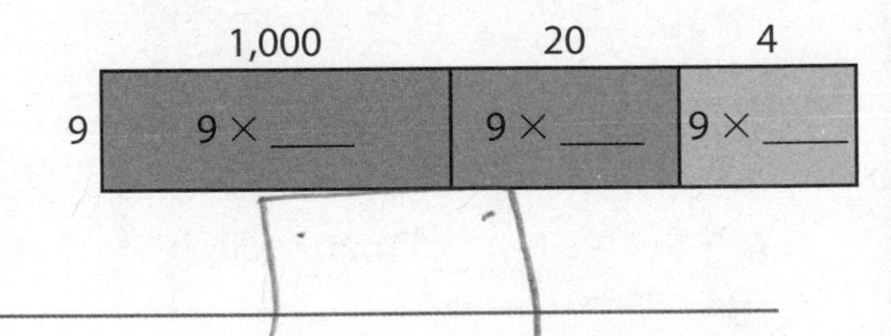

3. $43 \times 8 = (40 + 3) \times 8$

$= (\_\_\_ \times 8) + (\_\_\_ \times 8)$

$= \_\_\_ + \_\_\_$

$= \_\_\_$

4. $4 \times 742 = 4 \times (\_\_\_ + \_\_\_ + \_\_\_)$

$= (4 \times \_\_\_) + (4 \times \_\_\_) + (4 \times \_\_\_)$

$= \_\_\_ + \_\_\_ + \_\_\_$

$= \_\_\_$

5. $3 \times 482 =$ ____

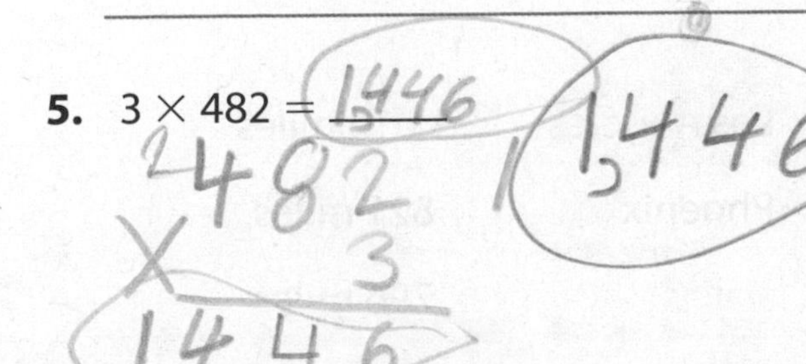

6. $4{,}591 \times 6 =$ ____

7. Each lobe on a Venus flytrap has 6 trigger hairs that sense and capture insects. A Venus flytrap has 128 lobes. How many trigger hairs does it have?

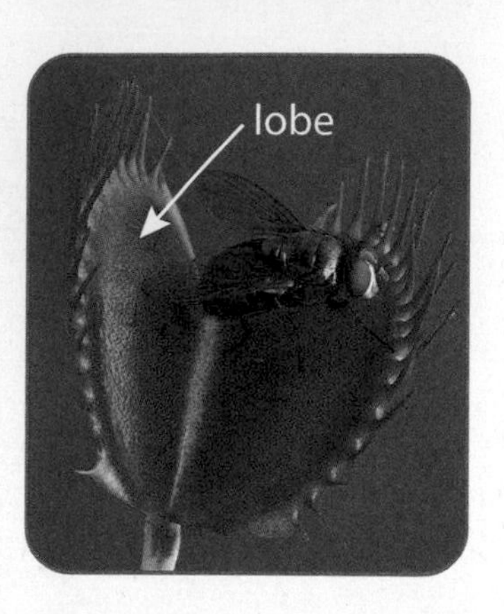

8. A human skeleton has 206 bones. How many bones do 4 skeletons have in all?

9. **Writing** Explain how you can find $5 \times 7{,}303$ using expanded form.

10. **MP Structure** Rewrite the expression as a product of two factors.

$(6{,}000 \times 3) + (70 \times 3) + (4 \times 3)$

11. **Modeling Real Life** A summer camp has 8 different groups of students. There are 35 students in each group. There are 24 shirts in a box. Will 9 boxes be enough for each student to get one shirt?

12. **Modeling Real Life** Firefighters respond to 65 calls in 1 week. Police officers respond to 8 times as many calls as firefighters in the same week. How many more calls do police officers respond to than firefighters in that week?

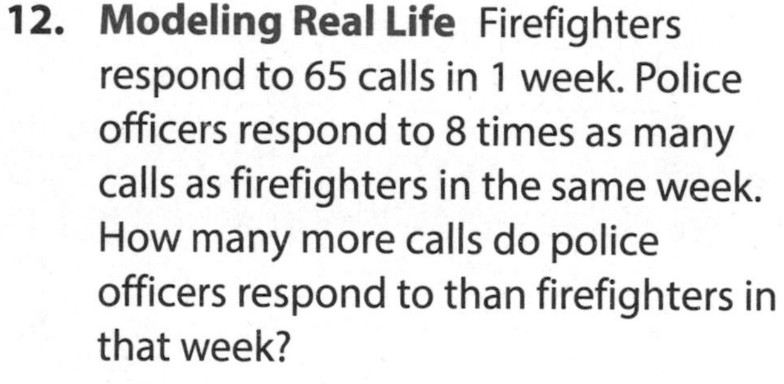

Review & Refresh

Write the total mass shown.

13.

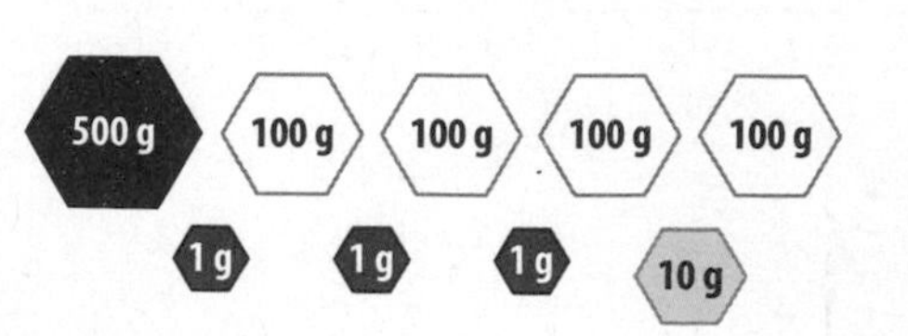

14.

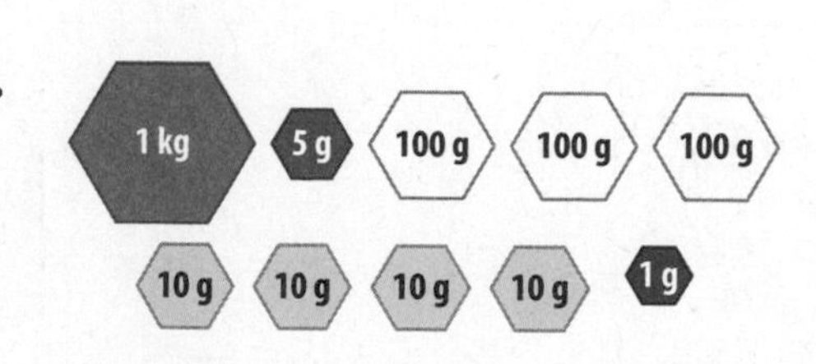

Name ______________________________

Use Partial Products to Multiply 3.6

Learning Target: Use place value and partial products to multiply.

Success Criteria:

- I can use place value to tell the value of each digit in a number.
- I can write the partial products for a multiplication problem.
- I can add the partial products to find a product.

Explore and Grow

Use the area model to find 263×4.

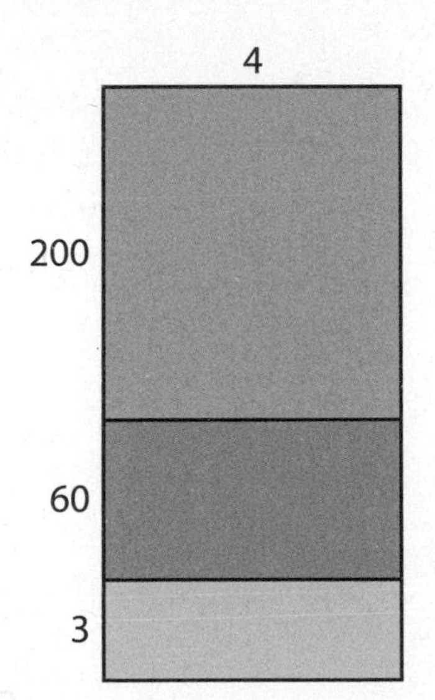

$$\begin{array}{r} 263 \\ \times\ 4 \\ \hline \square \\ \square \\ +\ \square \\ \hline \end{array}$$

Repeated Reasoning Does the product change if you multiply the ones first, then the tens and hundreds? Explain.

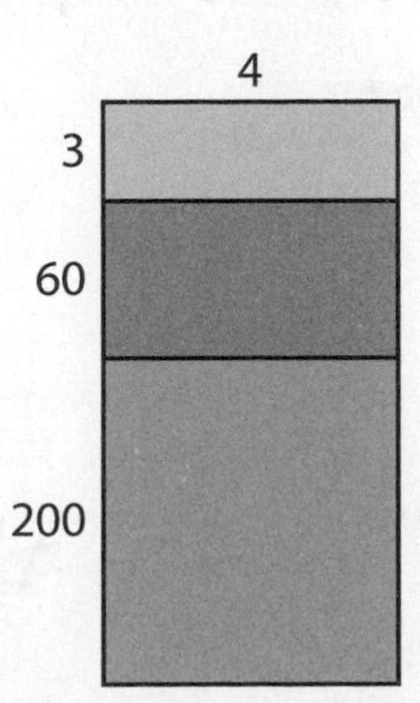

Think and Grow: Use Partial Products to Multiply

Partial products are found by breaking apart a factor into ones, tens, hundreds, and so on, and multiplying each of these by the other factor.

Example Use an area model and partial products to find 194 × 3.

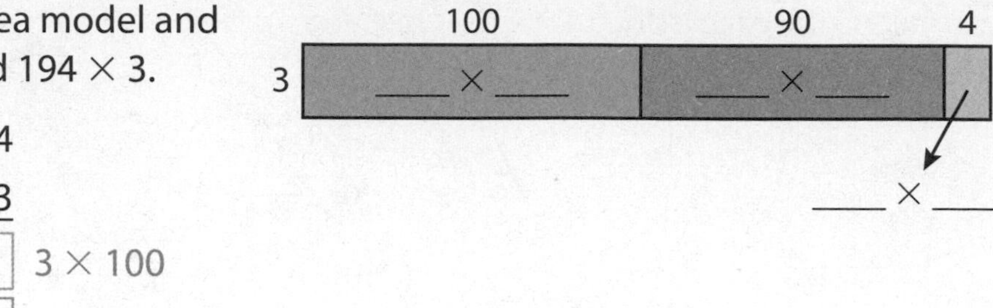

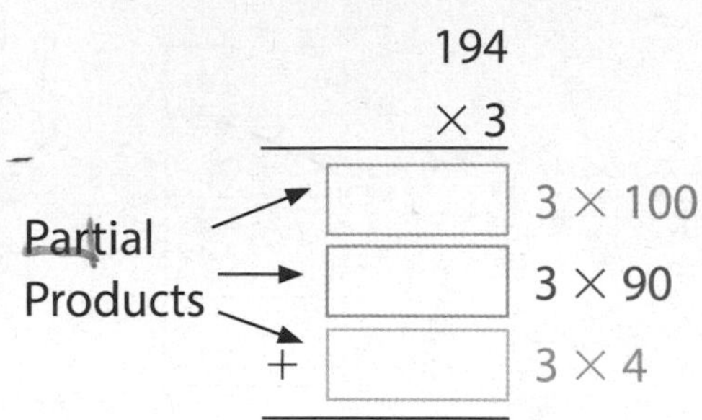

_____ Add the partial products.

So, 194 × 3 = _____.

Example Use place value and partial products to find 3,190 × 2.

You can find partial products in any order.

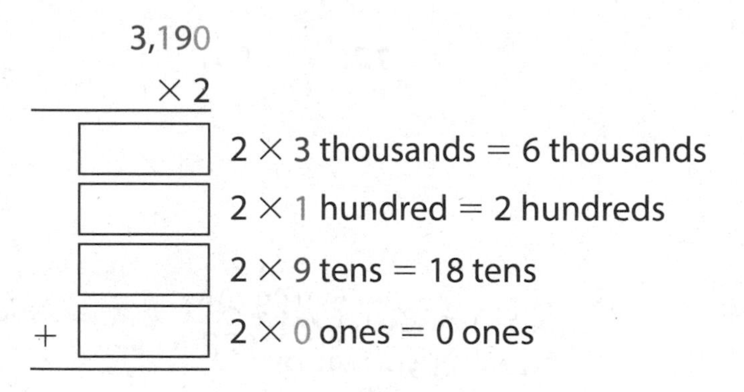

_____ Add the partial products.

So, 3,190 × 2 = _____.

Show and Grow *I can do it!*

Find the product.

1. 86 × 5

30

+ 400

430

2. 502 × 7

3. 5,367 × 4

Name ___

Apply and Grow: Practice

Find the product.

4. 27 × 6

42
+ 120
162

5. 493 × 9

27
810
+ 3,600
4,437

6. 6,982 × 8

16
640
7,200
+ 48,000
55,856

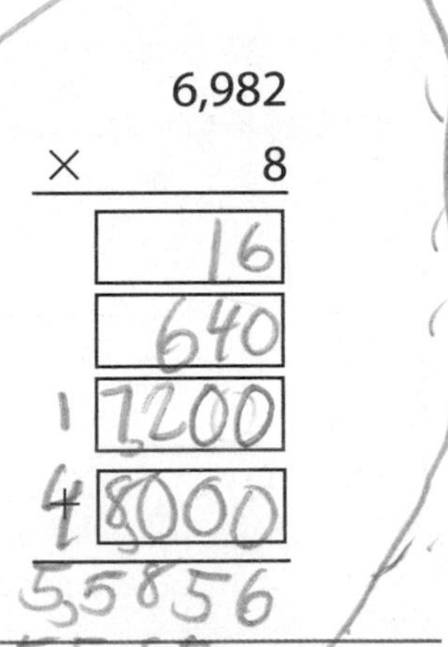

7. 69 × 2

8. 817 × 3

9. 3,962 × 5

10. 728 × 7

11. 1,036 × 9

12. 9,951 × 4

13. **YOU BE THE TEACHER** Descartes finds 472 × 3. Is he correct? Explain.

```
  472
×   3
 1200
  210
+ 6
 9,300
```

14. **DIG DEEPER!** Write the multiplication equation shown by the model.

4,000	320	72

97

15. **Writing** Write a multiplication word problem using the numbers 506 and 8. Then solve.

100

Think and Grow: Modeling Real Life

Example The Grand Prismatic Spring at Yellowstone National Park is 121 feet deep. Its width is 7 feet more than 3 times its depth. What is the width of the spring?

Multiply to find 3 times the depth.

$$\begin{array}{r} 121 \\ \times \quad 3 \\ \hline \square \\ \square \\ + \ \square \\ \hline \end{array}$$

_____ feet

Add to find 7 feet more.

_____ + 7 = _____ feet

So, the width of the Grand Prismatic Spring is _____ feet.

Show and Grow *I can think deeper!*

16. The mass of a giant squid is 202 kilograms. The mass of a beluga whale is 78 kilograms less than 6 times the mass of the giant squid. What is the mass of beluga whale?

17. Newton replaces all 4 tires on his car and pays $159 for each tire. Descartes replaces all 4 tires on his truck and pays $227 for each tire. How much more does Descartes pay to replace his tires?

18. There are 52 weeks and 1 day in a year. There are 52 weeks and 2 days in a leap year. How many weeks are there in 6 years if one of the years is a leap year?

Name ____________________

Multiply Two-Digit Numbers by One-Digit Numbers

3.7

Learning Target: Multiply two-digit numbers by one-digit numbers.

Success Criteria:

- I can multiply to find the partial products.
- I can show 10 ones regrouped as 1 ten.
- I can find the product.

Explore and Grow

Model 24×3. Draw to show your model. Then find the product.

$24 \times 3 =$ _____

How did you use regrouping to find the product?

Precision Use a model to find 15×8.

Think and Grow: Use Regrouping to Multiply

Example Find 32 × 6.

Estimate: 30 × 6 = ______

Step 1: Multiply the ones. Regroup.

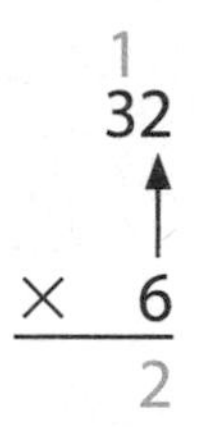

- 6 × 2 ones = 12 ones
- Regroup 12 ones as 1 ten and 2 ones.

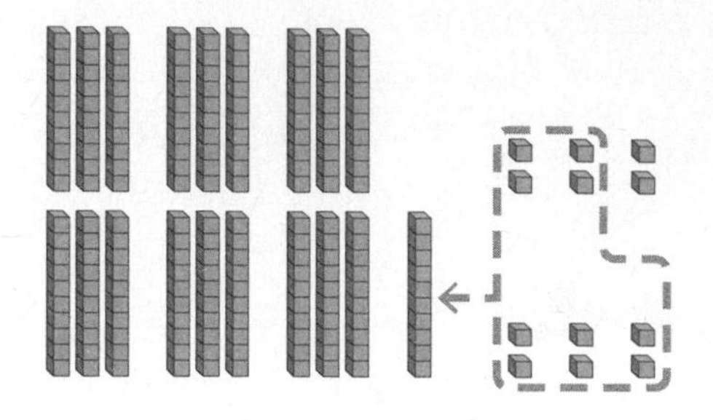

Step 2: Multiply the tens. Add any regrouped tens.

$$\begin{array}{r} ^{1} \\ 32 \\ \times\ \ 6 \\ \hline 192 \end{array}$$

- 6 × 3 tens = 18 tens
- 18 tens + 1 ten = 19 tens, or 1 hundred and 9 tens

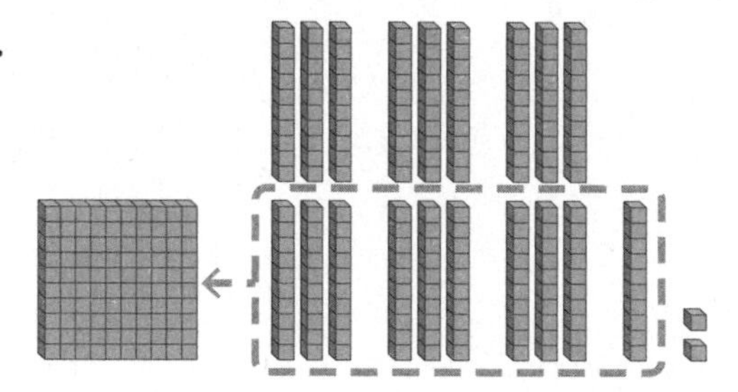

So, 32 × 6 = ______.

Check: Because ______ is close to the estimate, ______, the answer is reasonable.

Show and Grow *I can do it!*

1. Use the model to find the product.

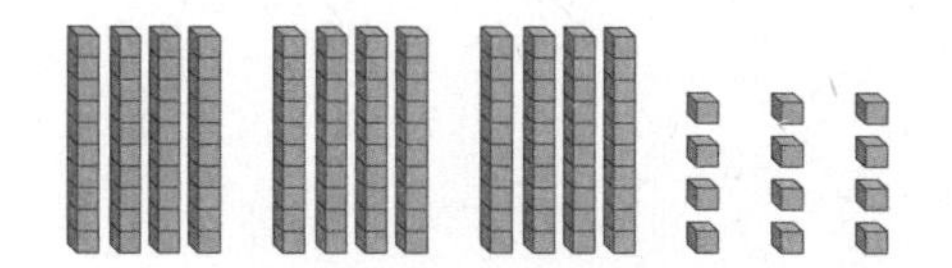

3 × 44 = ______

Find the product. Check whether your answer is reasonable.

2. Estimate: ______

$$\begin{array}{r} \square \\ 1\ 2 \\ \times\ \ \ 8 \\ \hline \end{array}$$

3. Estimate: ______

$$\begin{array}{r} \square \\ 4\ 7 \\ \times\ \ \ 5 \\ \hline \end{array}$$

4. Estimate: ______

$$\begin{array}{r} \square \\ 2\ 3 \\ \times\ \ \ 6 \\ \hline \end{array}$$

Name ____________________

Apply and Grow: Practice

Find the product. Check whether your answer is reasonable.

5. Estimate: _____ 25×7	**6.** Estimate: _____ 34×2	**7.** Estimate: _____ 56×9
8. Estimate: _____ 41×8	**9.** Estimate: _____ 73×4	**10.** Estimate: _____ 89×6
11. Estimate: _____ $65 \times 7 =$ _____	**12.** Estimate: _____ $3 \times 92 =$ _____	**13.** Estimate: _____ $47 \times 5 =$ _____

14. There are 8 questions during the first round of a game show. Each question is worth 15 points. What is the greatest number of points that a contestant can earn during the first round?

15. **DIG DEEPER!** Find the missing digits.

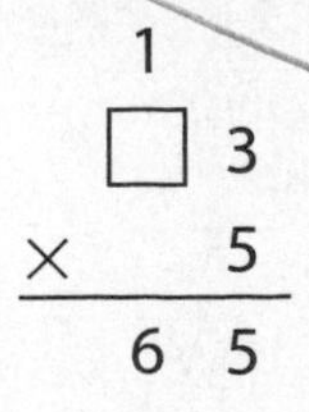

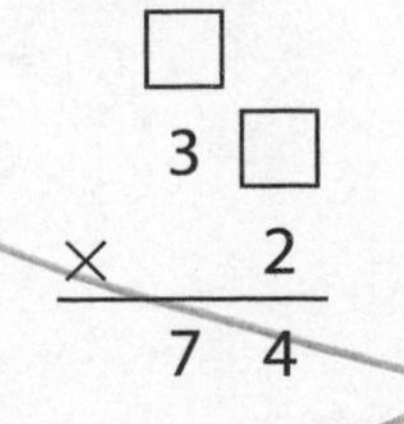

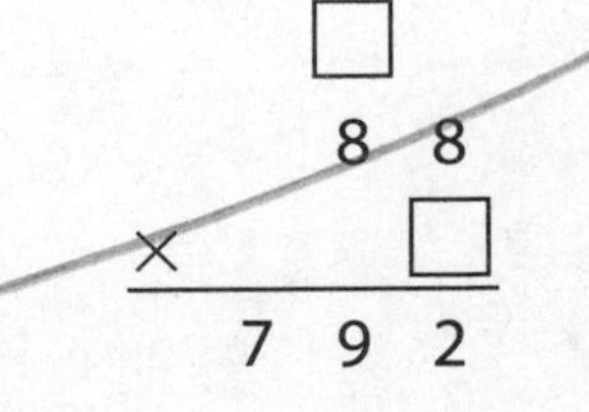

Think and Grow: Modeling Real Life

Example The fastest human on Earth can run up to 27 miles per hour. A pronghorn antelope can run up to 2 times as fast as the fastest human. A cheetah can run up to 61 miles per hour. Which animal can run faster?

Find how fast a pronghorn antelope can run.

$$\begin{array}{r} 27 \\ \times \quad 2 \\ \hline \end{array}$$

A pronghorn antelope can run up to ______ miles per hour.

Compare the fastest speeds of a pronghorn antelope and a cheetah.

So, a ________________ can run faster.

Show and Grow *I can think deeper!*

16. A band's goal is to produce 100 songs. The band has produced 6 albums with 13 songs on each album. Has the band reached its goal?

17. A teenager must practice driving with an adult for 50 hours before taking a driver's license test. A teenager practices driving with an adult 4 hours each week for 14 weeks. Has the teenager practiced long enough to take the test?

18. How much more does it cost to rent a personal watercraft for 3 hours than a motor boat for 3 hours?

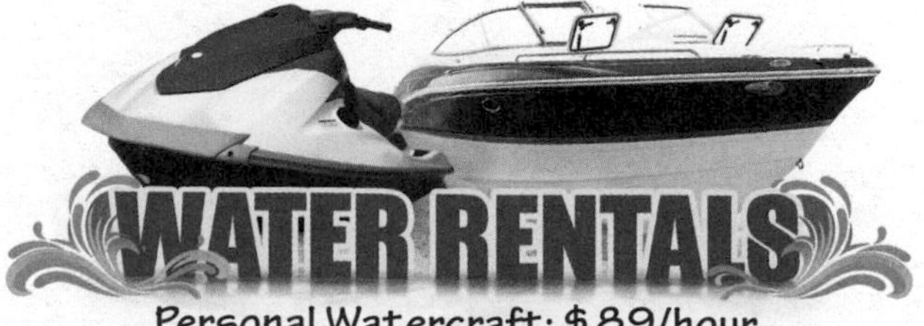

Personal Watercraft: $89/hour
Motor boat: $69/hour

Name ____________________

Learning Target: Multiply two-digit numbers by one-digit numbers.

Example Find 64×5.

Estimate: $60 \times 5 =$ 300

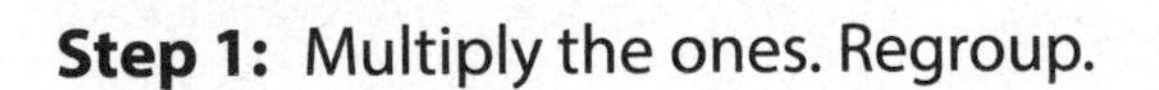

Step 1: Multiply the ones. Regroup.

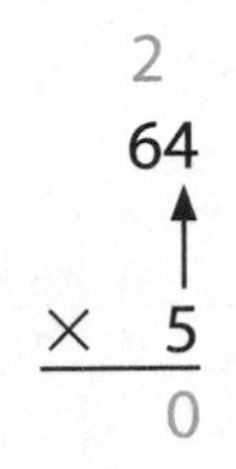

$$\begin{array}{r} {}^{2}\\ 64 \\ \times \quad 5 \\ \hline 0 \end{array}$$

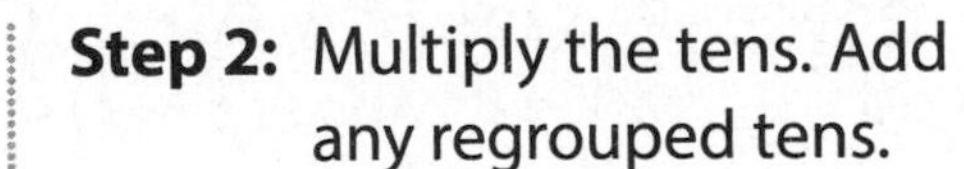

Step 2: Multiply the tens. Add any regrouped tens.

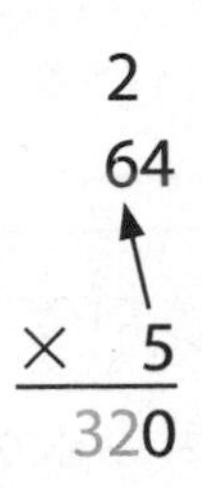

$$\begin{array}{r} {}^{2}\\ 64 \\ \times \quad 5 \\ \hline 320 \end{array}$$

So, $64 \times 5 =$ 320.

Check: Because 320 is close to the estimate, 300, the answer is reasonable.

1. Use the model to find the product.

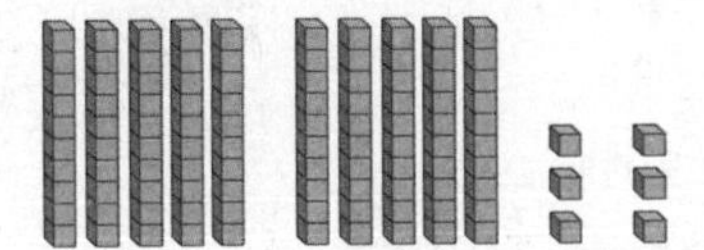

$2 \times 53 =$ ______

Find the product. Check whether your answer is reasonable.

2. Estimate: ______

$$\begin{array}{r} \square \\ 2\ \ 4 \\ \times \quad\ 3 \\ \hline \end{array}$$

3. Estimate: ______

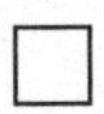

$$\begin{array}{r} \square \\ 1\ \ 8 \\ \times \quad\ 7 \\ \hline \end{array}$$

4. Estimate: ______

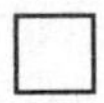

$$\begin{array}{r} \square \\ 3\ \ 3 \\ \times \quad\ 6 \\ \hline \end{array}$$

5. Estimate: ______

$$\begin{array}{r} 47 \\ \times \ 9 \\ \hline \end{array}$$

6. Estimate: ______

$$\begin{array}{r} 65 \\ \times \ 4 \\ \hline \end{array}$$

7. Estimate: ______

$$\begin{array}{r} 72 \\ \times \ 8 \\ \hline \end{array}$$

Find the product. Check whether your answer is reasonable.

8. Estimate: _____

$49 \times 5 =$ _____

9. Estimate: _____

$7 \times 86 =$ _____

10. Estimate: _____

$93 \times 3 =$ _____

11. You read 56 pages each week. How many pages do you read in 8 weeks?

12. MP **Number Sense** The sum of two numbers is 20. The product of the two numbers is 51. What are the two numbers?

13. MP **Reasoning** Your friend multiplies 58 by 6 and says that the product is 3,048. Is your friend's answer reasonable? Explain.

14. DIG DEEPER! How much greater is 4×26 than 3×26? Explain how you know without multiplying.

15. **Modeling Real Life** A self-balancing scooter travels 12 miles per hour. An all-terrain vehicle can travel 6 times as fast as the scooter. A go-kart can travel 67 miles per hour. Which vehicle can travel the fastest?

Self-balancing scooter

16. **Modeling Real Life** It takes a spaceship 3 days to reach the moon from Earth. It takes a spaceship 14 times as many days to reach Mars from Earth. How long would it take the spaceship to travel from Earth to Mars and back?

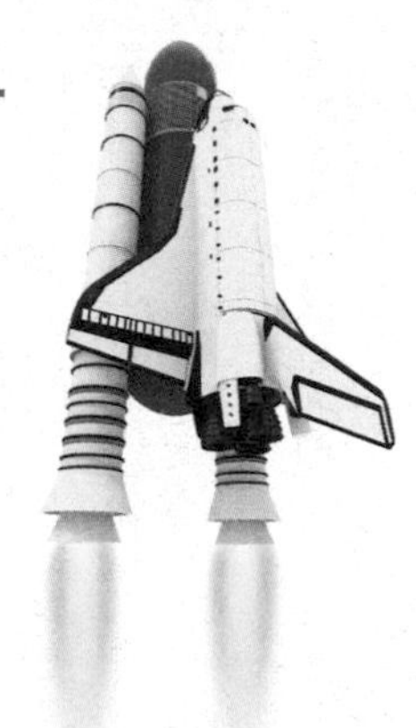

Review & Refresh

17. You spend 21 fewer minutes riding the bus to school than getting ready in the morning. You take 36 minutes to get ready. How long is the bus ride?

Name ______________________

Multiply Three- and Four-Digit Numbers by One-Digit Numbers

3.8

Learning Target: Multiply multi-digit numbers by one-digit numbers.

Success Criteria:

- I can multiply to find the partial products.
- I can show how to regroup more than 10 tens.
- I can find the product.

Explore and Grow

Use any strategy to find each product.

$7 \times 39 =$ _____

$7 \times 439 =$ _____

Structure How are the equations the same? How are they different?

Think and Grow: Use Regrouping to Multiply

Example Find 795×4.

Step 1: Multiply the ones. Regroup.

$$\begin{array}{r} ^{2} \\ 795 \\ \times \quad 4 \\ \hline 0 \end{array}$$

- 4×5 ones $= 20$ ones
- Regroup 20 ones as 2 tens and 0 ones.

Step 2: Multiply the tens. Add any regrouped tens.

$$\begin{array}{r} ^{32} \\ 795 \\ \times \quad 4 \\ \hline 80 \end{array}$$

- 4×9 tens $= 36$ tens
- 36 tens + 2 tens = 38 tens; Regroup 38 tens as 3 hundreds and 8 tens.

Step 3: Multiply the hundreds. Add any regrouped hundreds.

$$\begin{array}{r} ^{32} \\ 795 \\ \times \quad 4 \\ \hline 3{,}180 \end{array}$$

- 4×7 hundreds $= 28$ hundreds
- 28 hundreds + 3 hundreds = 31 hundreds, or 3 thousands and 1 hundred

So, $795 \times 4 =$ ______.

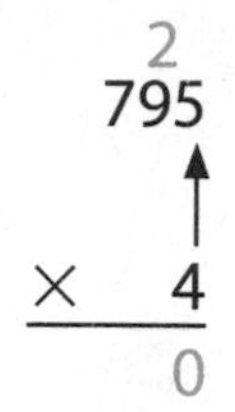

Example Find $6{,}084 \times 2$.

Estimate: $6{,}000 \times 2 =$ ______

$$\begin{array}{r} 6{,}084 \\ \times \quad 2 \\ \hline \end{array}$$

Multiply the ones, tens, hundreds, and thousands by 2. Regroup as necessary.

Check: Because ______ is close to the estimate, ______, the answer is reasonable.

Show and Grow *I can do it!*

Find the product. Check whether your answer is reasonable.

1. Estimate: ______

$$\begin{array}{r} 123 \\ \times \quad 5 \\ \hline \end{array}$$

2. Estimate: ______

$$\begin{array}{r} 907 \\ \times \quad 3 \\ \hline \end{array}$$

3. Estimate: ______

$$\begin{array}{r} 7{,}315 \\ \times \quad 6 \\ \hline \end{array}$$

Name ____________________

Apply and Grow: Practice

Find the product. Check whether your answer is reasonable.

4. Estimate: ______

$$\begin{array}{r} 120 \\ \times \quad 7 \\ \hline \end{array}$$

5. Estimate: ______

$$\begin{array}{r} 2{,}534 \\ \times \quad 4 \\ \hline \end{array}$$

6. Estimate: ______

$$\begin{array}{r} 7{,}617 \\ \times \quad 2 \\ \hline \end{array}$$

7. Estimate: ______

$$\begin{array}{r} 458 \\ \times \quad 8 \\ \hline \end{array}$$

8. Estimate: ______

$$\begin{array}{r} 8{,}823 \\ \times \quad 9 \\ \hline \end{array}$$

9. Estimate: ______

$$\begin{array}{r} 975 \\ \times \quad 6 \\ \hline \end{array}$$

10. Estimate: ______

$1{,}762 \times 3 =$ ______

11. Estimate: ______

$5 \times 5{,}492 =$ ______

12. Estimate: ______

$347 \times 7 =$ ______

Compare.

13. $6 \times 2{,}843$ ◯ $8 \times 1{,}645$

14. $6{,}582 \times 3$ ◯ $2{,}394 \times 8$

15. A roller coaster is twice as tall as a Ferris wheel that is 228 feet tall. How tall is the roller coaster?

16. **DIG DEEPER!** How can you estimate the number of digits in the product of 7 and 8,348?

Think and Grow: Modeling Real Life

Example The lengths of time that a penguin and an elephant seal can hold their breaths are shown. A sea turtle can hold its breath 7 times as long as a penguin. Which animal can hold its breath the longest?

Penguin:
1,200 seconds

Multiply to find how long a sea turtle can hold its breath.

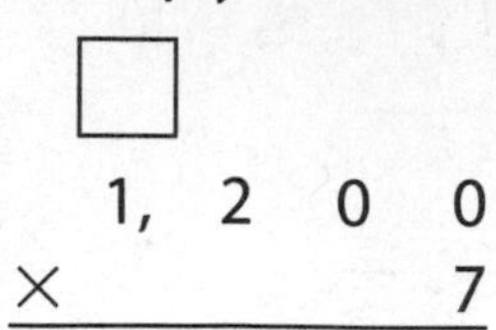

$$\begin{array}{r} 1,200 \\ \times \quad 7 \\ \hline \end{array}$$

Elephant Seal:
7,200 seconds

Compare the lengths of time that each animal can hold its breath.

The ____________ can hold its breath the longest.

Show and Grow *I can think deeper!*

17. You have 796 baseball cards and 284 hockey cards. You have 3 times as many football cards as hockey cards. Which type of card do you have the greatest number of?

18. A principal has $2,000 to spend on updating some of your school's tablets. She buys 4 tablets that each cost $299. How much money does the principal have left?

19. A train ticket from New York City to Miami costs $152. A train ticket from New York City to Orlando costs $144. A group of 8 friends is in New York City. How much money can the group save by going to Orlando instead of going to Miami?

Name ______________________

Homework & Practice 3.8

Learning Target: Multiply multi-digit numbers by one-digit numbers.

Example Find $4{,}932 \times 6$.

Estimate: $5{,}000 \times 6 =$ 30,000

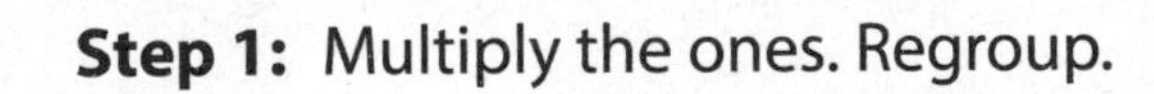

Step 1: Multiply the ones. Regroup.

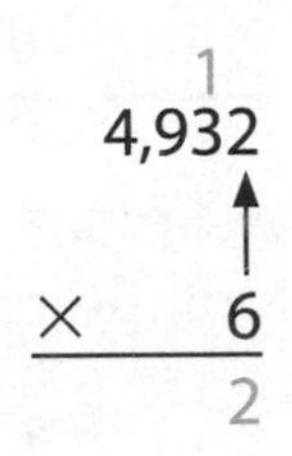

$$\begin{array}{r} {}^{1} \\ 4{,}932 \\ \times \quad 6 \\ \hline 2 \end{array}$$

Step 2: Multiply the tens. Add any regrouped tens.

$$\begin{array}{r} {}^{11} \\ 4{,}932 \\ \times \quad 6 \\ \hline 92 \end{array}$$

Step 3: Multiply the hundreds. Add any regrouped hundreds.

$$\begin{array}{r} {}^{5\,11} \\ 4{,}932 \\ \times \quad 6 \\ \hline 592 \end{array}$$

Step 4: Multiply the thousands. Add any regrouped thousands.

$$\begin{array}{r} {}^{5\,11} \\ 4{,}932 \\ \times \quad 6 \\ \hline 29{,}592 \end{array}$$

So, $4{,}932 \times 6 =$ 29,592.

Check: Because 29,592 is close to the estimate, 30,000, the answer is reasonable.

Find the product. Check whether your answer is reasonable.

1. Estimate: ______

$$\begin{array}{r} 304 \\ \times \quad 9 \\ \hline \end{array}$$

2. Estimate: ______

$$\begin{array}{r} 617 \\ \times \quad 8 \\ \hline \end{array}$$

3. Estimate: ______

$$\begin{array}{r} 5{,}939 \\ \times \quad 4 \\ \hline \end{array}$$

4. Estimate: ______

$$\begin{array}{r} 2{,}754 \\ \times \quad 3 \\ \hline \end{array}$$

5. Estimate: ______

$$\begin{array}{r} 8{,}465 \\ \times \quad 6 \\ \hline \end{array}$$

6. Estimate: ______

$$\begin{array}{r} 822 \\ \times \quad 7 \\ \hline \end{array}$$

Find the product. Check whether your answer is reasonable.

7. Estimate: ______

$629 \times 5 =$ ______

8. Estimate: ______

$7 \times 1{,}836 =$ ______

9. Estimate: ______

$453 \times 3 =$ ______

Compare.

10. $3 \times 6{,}782$ ◯ $2 \times 3{,}391$

11. $1{,}392 \times 9$ ◯ $2{,}493 \times 4$

12. A backpacker hikes the Buckeye Trail 5 times. The trail is 1,444 miles long. How many miles has he hiked on the Buckeye Trail in all?

13. MP **Number Sense** What number is 980 more than the product of 6,029 and 8?

14. **YOU BE THE TEACHER** Newton says that the product of a three-digit number and a one-digit number is always a three-digit number. Is Newton correct? Explain.

15. **Modeling Real Life** The numbers of songs Newton and Descartes download are shown. You download 2 times as many songs as Descartes. Who downloads the most songs?

Newton: 351 songs

Descartes: 167 songs

Review & Refresh

Find the sum or difference. Use the inverse operation to check.

16. $\begin{array}{r} 847 \\ -\ 162 \\ \hline \end{array}$

17. $\begin{array}{r} 612 \\ +\ 289 \\ \hline \end{array}$

18. $\begin{array}{r} 500 \\ -\ 351 \\ \hline \end{array}$

Name ______________________________

Use Properties to Multiply 3.9

Learning Target: Use properties to multiply.

Success Criteria:

- I can use the Commutative Property of Multiplication to multiply.
- I can use the Associative Property of Multiplication to multiply.
- I can use the Distributive Property to multiply.

Use any strategy to find each product. Explain the strategy you used to find each product.

$7 \times 25 \times 4 =$ _____	$5 \times 79 \times 2 =$ _____
$9 \times 5{,}001 =$ _____	$9 \times 4{,}999 =$ _____

Construct Arguments Compare your strategies with your partner's. How are they alike? How are they different?

Think and Grow: Use Properties to Multiply

You can use properties to multiply.

Example Find 8×250.

$8 \times 250 = (4 \times 2) \times 250$ Think: $8 = 4 \times 2$

$= 4 \times (2 \times 250)$ Associative Property of Multiplication

$= 4 \times$ ______

$=$ ______ So, $8 \times 250 =$ ______.

Example Find 5×698.

$5 \times 698 = 5 \times (700 - 2)$ Think: $698 = 700 - 2$

$= (5 \times 700) - (5 \times 2)$ Distributive Property

$=$ ______ $-$ ______

$=$ ______ So, $5 \times 698 =$ ______.

Example Find $4 \times 9 \times 25$.

$4 \times 9 \times 25 = 9 \times 4 \times 25$ Commutative Property of Multiplication

$= 9 \times$ ______

$=$ ______ So, $4 \times 9 \times 25 =$ ______.

Show and Grow *I can do it!*

Use properties to find the product. Explain your reasoning.

1. 6×150

2. 3×494

3. $25 \times 7 \times 4$

Name ______________________________

Apply and Grow: Practice

Use properties to find the product. Explain your reasoning.

4. 7 × 798

5. 350 × 6

6. 106 × 5

7. 4 × 625

8. 395 × 8

9. 2 × 7 × 15

10. 430 × 2

11. 8 × 150

12. 3 × 1,997

13. 25 × 9 × 2

14. 404 × 6

15. 4 × 2,004

16. **Which One Doesn't Belong?** Which expression does *not* belong with the other three?

(3 × 30) + (3 × 7)   (3 × 40) − (3 × 3)

3 × (30 + 7)   3 × 3 × 7

17. MP **Number Sense** Use properties to find each product.

9 × 80 = 720, so 18 × 40 = ______.

5 × 70 = 350, so 5 × 72 = ______.

Think and Grow: Modeling Real Life

Top speed:
120 miles per hour

Example The fastest recorded speed of a dragster car in the United States was 31 miles per hour less than 3 times the top speed of the roller coaster. What was the speed of the car?

Multiply to find 3 times the top speed of the roller coaster.

$3 \times 120 = 3 \times (100 + 20)$ Rewrite 120 as 100 + 20.

$= (3 \times 100) + (3 \times 20)$ Distributive Property

$=$ _____ + _____

$=$ _____ miles per hour

Subtract to find 31 miles per hour less.

_____ − 31 = _____ miles per hour

So, the speed of the car was _____ miles per hour.

Show and Grow *I can think deeper!*

18. In 2016, a theme park used 300 drones for a holiday show. In 2017, China used 200 fewer than 4 times as many drones for a lantern festival. How many drones did China use?

19. A subway train has 8 cars. Each car can hold 198 passengers. How many passengers can two subway trains hold?

20. You plant cucumbers, green beans, squash, and corn in a community garden. You plant 3 rows of each vegetable with 24 seeds in each row. How many seeds do you plant?

Name ______________________________

Homework & Practice 3.9

Learning Target: Use properties to multiply.

You can use properties to multiply.

Example Find 7×496.

$7 \times 496 = 7 \times (500 - 4)$ Think: $496 = 500 - 4$

$= (7 \times 500) - (7 \times 4)$ Distributive Property

$= \underline{3{,}500} - \underline{28}$

$= \underline{3{,}472}$

So, $7 \times 496 = \underline{3{,}472}$.

Example Find $2 \times 5 \times 25$.

$2 \times 5 \times 25 = 5 \times 2 \times 25$ Commutative Property of Multiplication

$= 5 \times \underline{50}$

$= \underline{250}$

So, $2 \times 5 \times 25 = \underline{250}$.

Use properties to find the product. Explain your reasoning.

1. 3×497	**2.** 36×9	**3.** 8×350
4. $25 \times 8 \times 4$	**5.** 999×5	**6.** 9×402
7. 509×4	**8.** $2 \times 9 \times 15$	**9.** $3{,}998 \times 7$

10. **YOU BE THE TEACHER** Is Descartes correct? Explain.

$$\begin{aligned}895 \times 4 &= (900 - 5) \times 4 \\ &= (900 \times 4) + (5 \times 4) \\ &= 3{,}600 + 20 \\ &= 3{,}620\end{aligned}$$

11. **DIG DEEPER!** Complete the square so that the product of each row and each column is 2,400.

4		
200		2
		2

12. **Modeling Real Life** The height of the Great Pyramid of Giza is 275 feet shorter than 2 times the height of the Luxor Hotel in Las Vegas. How tall is the Great Pyramid of Giza?

365 ft

Luxor Hotel in Las Vegas

13. **Modeling Real Life** Some firefighters are testing their equipment. Water from their firetruck hose hits the wall 12 feet higher than 7 times where the spray from a fire extinguisher hits the wall. How many feet high does the water from the hose hit the wall?

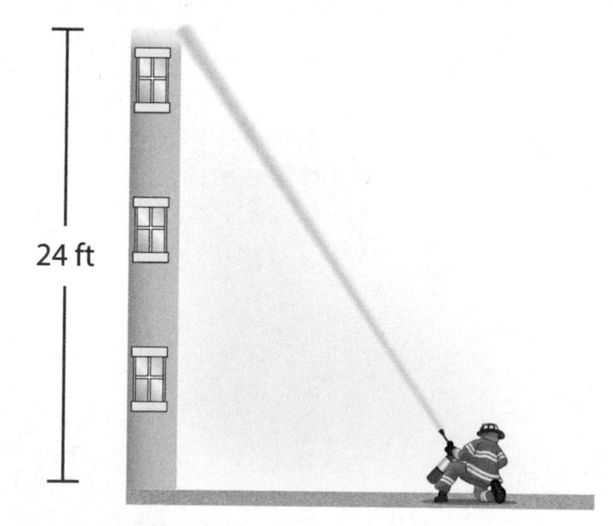

Review & Refresh

14. You want to learn 95 new vocabulary words. You learn 5 words the first week and an equal number of words each week for the next 9 weeks. How many words do you learn in each of the 9 weeks?

Name ______________________________

Problem Solving: Multiplication 3.10

Learning Target: Solve multi-step word problems involving multiplication.

Success Criteria:

- I can understand a problem.
- I can make a plan to solve using letters to represent the unknown numbers.
- I can solve a problem using an equation.

Explore and Grow

Make a plan to solve the problem.

A group of sea otters that swim together is called a raft. There are 37 otters in a raft. Each otter eats about 16 pounds of food each day. About how many pounds of food does the raft eat in 1 week?

Critique Reasoning Compare your plan to your partner's. How are your plans alike? How are they different?

Think and Grow: Problem Solving: Multiplication

Example A coach buys 6 cases of sports drinks and spends $60. Each case has 28 bottles. A team drinks 85 bottles at a tournament. How many bottles are left?

Understand the Problem

What do you know?

- The coach buys 6 cases.
- The coach spends $60.
- Each case has 28 bottles.
- The team drinks 85 bottles.

What do you need to find?

- You need to find how many bottles are left.

Make a Plan

How will you solve?

- Multiply 28 by 6 to find the total number of bottles in 6 cases.
- Then subtract 85 bottles from the product to find how many bottles are left.
- The amount of money the coach spends is unnecessary information.

Solve

Step 1: How many bottles are in 6 cases?

28	28	28	28	28	28

k

k is the unknown product.

$28 \times 6 = k$

☐

 2 8
× 6

$k =$ ______

Step 2: Use k to find how many bottles are left.

85	n

$k =$ ______

n is the unknown difference.

______ $- 85 = n$

☐
− 85

$n =$ ______

There are ______ bottles left.

Show and Grow I can do it!

1. Explain how you can check whether your answer above is reasonable.

Name ______________________________

Apply and Grow: Practice

Understand the problem. What do you know? What do you need to find? Explain.

2. A zookeeper has 4 boxes. There are 137 grams of leaves in each box. A koala eats 483 grams of leaves in 1 day. The zookeeper wants to know how many grams of leaves are left.

3. A beekeeper has 2 hives. Hive A produces 14 pounds of honey. Hive B produces 4 times as much honey as Hive A. The beekeper wants to know how many pounds of honey are produced in all.

Understand the problem. Then make a plan. How will you solve? Explain.

4. A runner completes 12 races each year. He improves his time by 10 seconds each year. Each race is 5 kilometers long. The runner wants to know how many kilometers he runs in races in 3 years.

5. A volunteer bikes 4 miles in all to travel from her home to a shelter and back. At the shelter, she walks a dog 1 mile. The volunteer wants to know how many miles she travels doing these tasks for 28 days.

6. Cats have 32 muscles in each ear. Humans have 12 ear muscles in all. How many more muscles do cats have in both ears than humans have in both ears?

7. A school has 5 hallways. Each hallway has 124 lockers. 310 lockers are red. 586 lockers are in use. How many lockers are *not* in use?

Think and Grow: Modeling Real Life

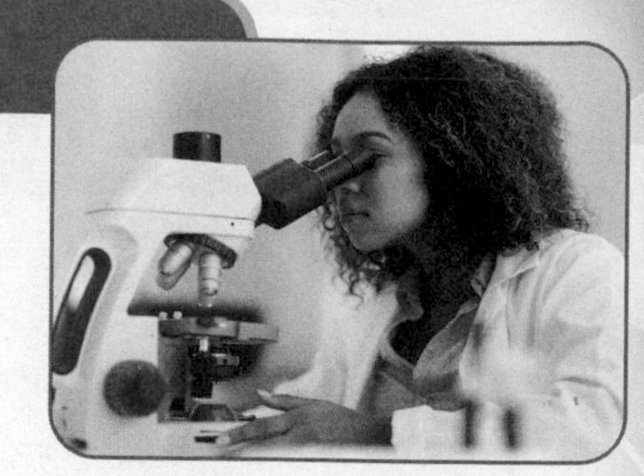

Example A group of scientists has $7,500 to spend on microscopes and balances. They buy 3 microscopes that each cost $1,642 and 6 balances that each cost $236. How much money do the scientists have left to spend?

Think: What do you know? What do you need to find? How will you solve?

Step 1: How much do the scientists spend on microscopes?

1,642	1,642	1,642

m

m is the unknown product.

$1{,}642 \times 3 = m$

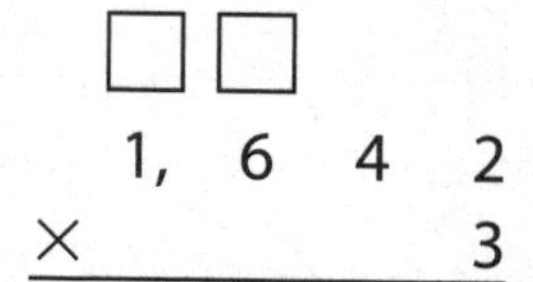

$m =$ ______

Step 2: How much do the scientists spend on balances?

236	236	236	236	236	236

r

r is the unknown product.

$236 \times 6 = r$

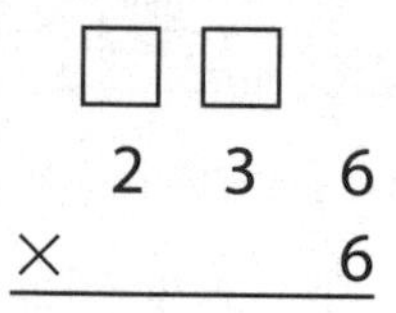

$r =$ ______

Step 3: Use m and r to find how much the scientists spend in all.

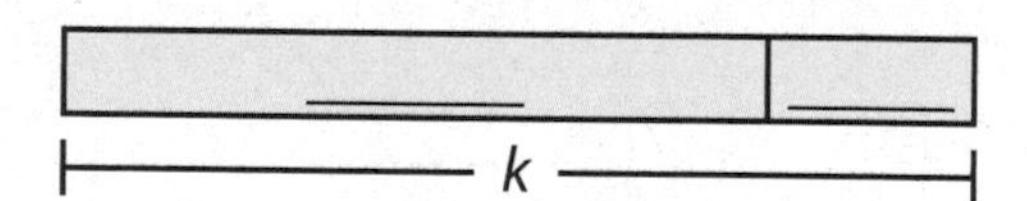

k is the unknown sum.

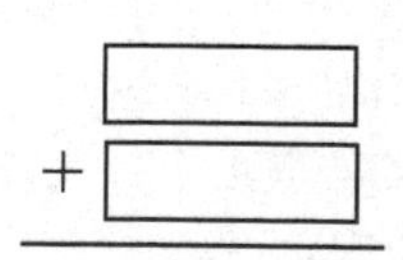

$k =$ ______

Step 4: Use k to find how much money the scientists have left.

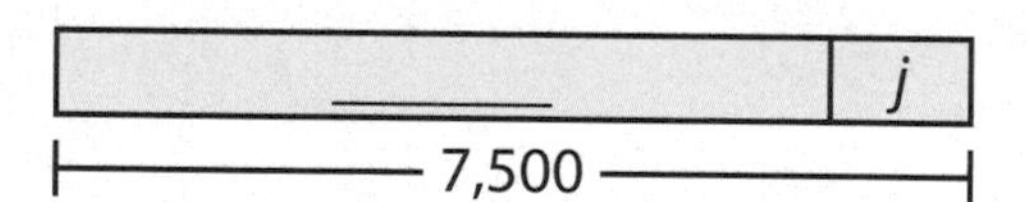

j is the unknown difference.

7,500
− ☐

$j =$ ______

The scientists have ____________ left to spend.

Show and Grow *I can think deeper!*

8. City A has 3,296 fourth graders. City B has 3 times as many fourth graders as City A. City C has 5 times as many fourth graders as City B. How many fourth graders are in all three cities?

Name ___________________________

Homework & Practice 3.10

Learning Target: Solve multi-step word problems involving multiplication.

Example You earn 206 points in Round 1 of a game. In Round 2, you earn 1,341 points. Then you find a magic star that makes your points from Round 2 worth 7 times as much. How many points do you earn in all?

Think: What do you know? What do you need to find? How will you solve?

Step 1: How many points do you earn from having star power?

1,341	1,341	1,341	1,341	1,341	1,341	1,341

p

p is the unknown product.

$1{,}341 \times 7 = p$

```
  2 2
 1, 3 4 1
×       7
---------
 9, 3 8 7
```

$p = \underline{9{,}387}$

Step 2: Use p to find how many points you earn in all.

206	$p = \underline{9{,}387}$

t

t is the unknown sum.

$206 + \underline{9{,}387} = t$

```
   206
+ 9,387
-------
  9,593
```

$t = \underline{9{,}593}$

You earn $\underline{9{,}593}$ points in all.

Understand the problem. Then make a plan. How will you solve? Explain.

1. *Raku* is a Japanese-inspired art form. An artist *fires*, or bakes, a raku pot at 1,409 degrees Fahrenheit. The artist fires a porcelain pot at 266 degrees Fahrenheit less than 2 times the temperature at which the raku pot is fired. You want to find the temperature at which the porcelain pot is fired.

2. You practice basketball 6 times each week. Each basketball practice is 55 minutes long. You practice dance for 225 minutes altogether each week. You want to find how many minutes you practice basketball and dance in all each week.

3. You buy 7 books of stamps. There are 35 stamps in each book. You give some away and have 124 stamps left. How many stamps did you give away?

4. Your neighbor fills his car's gasoline tank with 9 gallons of gasoline. Each gallon of gasoline allows him to drive 23 miles. Can he drive for 210 miles without filling his gasoline tank? Explain.

5. **Writing** Write and solve a two-step word problem that can be solved using multiplication as one step.

6. **Modeling Real Life** There are 1,203 pictures taken for a yearbook. There are 124 student pictures for each of the 6 grades. There are also 7 pictures for each of the 23 school clubs. The rest of the pictures are teacher or candid pictures. How many teacher or candid pictures are there?

7. **Modeling Real Life** A construction worker earns $19 each hour she works. A supervisor earns $35 each hour she works. How much money do the construction worker and supervisor earn in all after 8 hours of work?

8. **Modeling Real Life** A family of 8 has $1,934 to spend on a vacation that is 1,305 miles away. They buy a $217 plane ticket and an $8 shirt for each person. How much money does the family have left?

Review & Refresh

What fraction of the whole is shaded?

9.

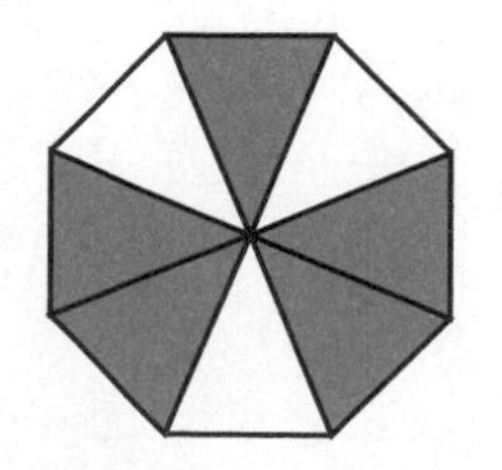

$\frac{\square}{\square}$ is shaded.

10.

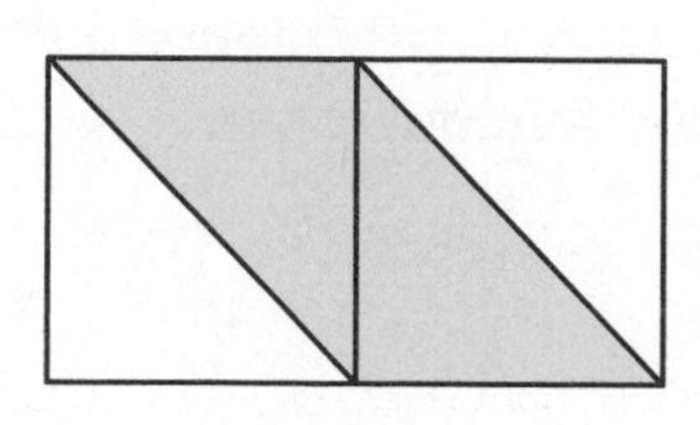

$\frac{\square}{\square}$ is shaded.

Name ______________________________

Performance Task 3

Sounds are vibrations that travel as waves through solids, liquids, and gases. Sound waves travel 1,125 feet per second through air.

1. You see a flash of lightning 5 seconds before you hear the thunder. How far away is the storm?

---

2. Sound waves travel 22,572 feet per second faster through iron than through diamond. The speed of sound through diamond is 39,370 feet per second.

a. Estimate the speed of sound through iron in feet per second.

b. What is the actual speed of sound through iron in feet per second?

c. Is your estimate close to the exact speed of sound through iron? Explain.

---

3. Sound waves travel about 4 times faster through water than through air.

a. What is the speed of sound through water in feet per second?

b. A horn blows under water. A diver is about 9,000 feet away from the horn. About how many seconds does it take the diver to hear the sound of the horn?

---

4. Do sound waves travel the fastest through solids, liquids, or gases? Explain.

Multiplication Quest

Directions:

1. Players take turns rolling a die. Players solve problems on their boards to race the knights to their castles.
2. On your turn, solve the next multiplication problem in the row of your roll.
3. The first player to get a knight to a castle wins!

Roll				
1	7×8	32×3	629×4	$5{,}107 \times 6$
2	3×9	56×4	248×1	$3{,}816 \times 8$
3	5×2	81×5	921×9	$7{,}249 \times 7$
4	4×6	90×9	455×8	$9{,}683 \times 2$
5	8×1	12×7	806×3	$4{,}749 \times 5$
6	6×7	79×2	573×6	$8{,}106 \times 4$

Name ______________________________

Chapter Practice 3

3.1 Understand Multiplicative Comparisons

Write two comparison sentences for the equation.

1. $72 = 8 \times 9$

2. $60 = 10 \times 6$

Write an equation for the comparison sentence.

3. 28 is 4 times as many as 7.

4. 40 is 8 times as many as 5.

5. Newton saves \$40. Descartes saves \$25 more than Newton. How much money does Descartes save?

6. Your friend is 10 years old. Your neighbor is 4 times as old as your friend. How old is your neighbor?

3.2 Multiply Tens, Hundreds, and Thousands

Find the product.

7. $6 \times 90 =$ ______

8. $3{,}000 \times 1 =$ ______

9. $4 \times 200 =$ ______

10. $4{,}000 \times 4 =$ ______

11. $8 \times 700 =$ ______

12. $2 \times 60 =$ ______

Find the missing factor.

13. ______ $\times 200 = 1{,}000$

14. $5 \times$ ______ $= 450$

15. ______ $\times 800 = 6{,}400$

3.3 Estimate Products by Rounding

Estimate the product.

16. 5×65

17. 2×903

18. $7 \times 3,592$

Find two estimates that the product is between.

19. 8×32

20. 4×284

21. $6 \times 5,945$

22. A charity organizer raises $9,154 each month for 6 months. To determine whether the charity raises $50,000, can you use an estimate, or is an exact answer required? Explain.

3.4 Use Distributive Property to Multiply

Find the product.

23. $3 \times 14 =$ _____

24. $18 \times 9 =$ _____

25. $36 \times 5 =$ _____

26. $8 \times 56 =$ _____

27. $6 \times 67 =$ _____

28. $83 \times 2 =$ _____

29. MP **Structure** Use the Distributive Property to write an equation shown by the model.

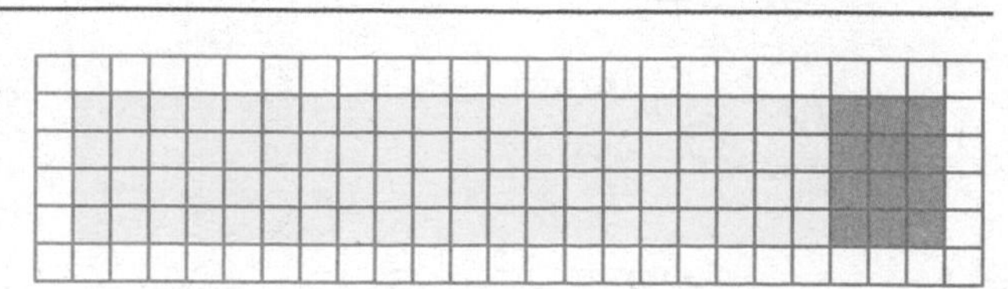

Use Expanded Form to Multiply

Find the product.

30. $487 \times 3 =$ ______

31. $5 \times 7{,}402 =$ ______

32. $8{,}395 \times 7 =$ ______

3.6 Use Partial Products to Multiply

Find the product.

33. $\begin{array}{r} 266 \\ \times \quad 9 \\ \hline \end{array}$

34. $\begin{array}{r} 85 \\ \times \ 8 \\ \hline \end{array}$

35. $\begin{array}{r} 7{,}032 \\ \times \quad 4 \\ \hline \end{array}$

36. MP **Number Sense** Which three numbers are the partial products that you add to find the product of 518 and 2?

100 200 16 1,000 14 20

Multiply Two-Digit Numbers by One-Digit Numbers

Find the product. Check whether your answer is reasonable.

37. Estimate: ______

$\begin{array}{r} 19 \\ \times \ 6 \\ \hline \end{array}$

38. Estimate: ______

$\begin{array}{r} 73 \\ \times \ 7 \\ \hline \end{array}$

39. Estimate: ______

$\begin{array}{r} 58 \\ \times \ 5 \\ \hline \end{array}$

3.8 Multiply Three- and Four-Digit Numbers by One-Digit Numbers

Find the product. Check whether your answer is reasonable.

40. Estimate: ______

$402 \times 3 =$ ______

41. Estimate: ______

$8 \times 3{,}861 =$ ______

42. Estimate: ______

$977 \times 2 =$ ______

Compare.

43. 308×6 ◯ 5×408

44. 6×789 ◯ $2 \times 2{,}367$

45. 454×3 ◯ 313×4

3.9 Use Properties to Multiply

Use properties to find the product. Explain your reasoning.

46. $25 \times 9 \times 4$

47. 8×250

48. 3×497

49. $2 \times 8 \times 15$

50. 699×9

51. $1{,}003 \times 6$

3.10 Problem Solving: Multiplication

52. A musical cast sells 1,761 tickets for a big show. The cast needs to complete 36 days of rehearsal. Each rehearsal is 8 hours long. The cast has rehearsed for 102 hours so far. How many hours does the cast have left to rehearse?

Name ____________________

Cumulative Practice 1–3

1. Which number is greater than 884,592?

Ⓐ 89,621

Ⓑ 884,592

Ⓒ 805,592

Ⓓ 894,592

---

2. What is the difference of 30,501 and 6,874?

---

3. A teenager will send about 37,000 text messages within 1 year. Which numbers could be the exact number of text messages sent?

☐ 37,461

☐ 36,834

☐ 37,050

☐ 37,503

☐ 36,006

☐ 36,752

---

4. What is the product of 4,582 and 6?

Ⓐ 27,492

Ⓑ 24,082

Ⓒ 27,432

Ⓓ 42,117

---

5. Which expression is equal to $246{,}951 + 73{,}084$?

Ⓐ $246{,}951 + 70{,}000 + 3{,}000 + 800 + 40$

Ⓑ $246{,}951 + 7{,}000 + 300 + 80 + 4$

Ⓒ $246{,}951 + 70{,}000 + 3{,}000 + 80 + 4$

Ⓓ $246{,}951 + 70{,}000 + 3{,}000 + 800 + 4$

6. Newton reads the number "four hundred six thousand, twenty-nine" in a book. What is this number written in standard form?

Ⓐ 4,629 Ⓑ 406,029

Ⓒ 460,029 Ⓓ 406,290

---

7. Which expressions have a product of 225?

☐ 9×25 ☐ 75×3

☐ 56×4 ☐ 5×45

---

8. What is the greatest possible number you can make with the number cards below?

6 1 8 4 2

Ⓐ 84,621 Ⓑ 86,421

Ⓒ 12,468 Ⓓ 68,421

---

9. What is the sum of 62,671 and 48,396?

Ⓐ 111,067 Ⓑ 11,067

Ⓒ 100,967 Ⓓ 100,067

10. Your friend drinks 8 glasses of water each day. He wants to know how many glasses of water he will drink in 1 year. Between which two estimates is the number of glasses he will drink in a year?

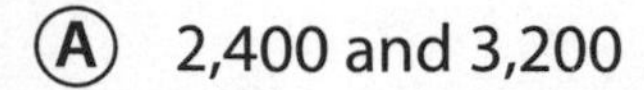

Ⓐ 2,400 and 3,200 Ⓑ 240 and 320

Ⓒ 3,200 and 4,000 Ⓓ 320 and 400

11. Which statement about the number 420,933 is true?

Ⓐ The value of the 2 is 2,000.

Ⓑ The 4 is in the ten thousands place.

Ⓒ The value of the 3 in the tens place is ten times the value of the 3 in the ones place.

Ⓓ There are 9 tens.

---

12. Descartes rounds to the nearest ten thousand and gets an estimate of 360,000. Which expressions could be the problem he estimated?

☐ 785,462 − 432,587

☐ 480,012 − 127,465

☐ 74,621 − 10,354

☐ 398,650 − 41,579

---

13. There are 8 students in a book club. There are 5 times as many students in a drama club as the book club. How many students are in the drama club?

Ⓐ 13 students

Ⓑ 45 students

Ⓒ 3 students

Ⓓ 40 students

---

14. The force required to shatter concrete is 3 times the amount of force required to shatter plexiglass. The force required to shatter plexiglass is 72 pounds per square foot. What is the force required to shatter concrete?

Ⓐ 216 pounds per square foot

Ⓑ 69 pounds per square foot

Ⓒ 75 pounds per square foot

Ⓓ 2,106 pounds per square foot

---

15. Which number, when rounded to the nearest hundred thousand, is equal to 100,000?

Ⓐ 9,802

Ⓑ 83,016

Ⓒ 152,853

Ⓓ 46,921

16. You want to find $5,193 \times 8$. Which expressions show how to use the Distributive Property to find the product?

- ◯ $8 \times (5,000 + 100 + 90 + 3)$
- ◯ $(8 \times 5,000) + (8 \times 100) + (8 \times 90) + (8 \times 3)$
- ◯ $40,000 + 800 + 720 + 24$
- ◯ $(8 \times 3) + (8 \times 9) + (8 \times 1) + (8 \times 5)$

---

17. Use the table to answer the questions.

Part A Five friends each buy all of the items listed in the table. Use rounding to find about how much money they spend in all.

Trampoline Park Prices

Item	Price
90-minute jump pass	\$21
Water Bottle	\$8
T-shirt	\$15

Part B An Ultimate Package costs \$40 and includes all of the items listed in the table. About how much money would the 5 friends have saved in all if they would have bought an Ultimate Package instead of purchasing the items individually? Explain.

---

18. A pair of walruses weigh 4,710 pounds together. The female weighs 1,942 pounds. How much more does the male weigh than the female?

Ⓐ 826 pounds

Ⓑ 6,652 pounds

Ⓒ 2,768 pounds

Ⓓ 2,710 pounds

---

19. You use compensation as shown. What is the final step?

$$\begin{array}{r} 6,325 \\ -\ 2,258 \\ \hline \end{array} \; - 58 \longrightarrow \begin{array}{r} 6,325 \\ -\ 2,200 \\ \hline 4,125 \end{array}$$

Ⓐ Add 58 to 4,125.

Ⓑ Subtract 58 from 4,125.

Ⓒ Subtract 4,125 from 6,325.

Ⓓ There is no final step. You are finished.

Name ______________________________

STEAM Performance Task 1–3

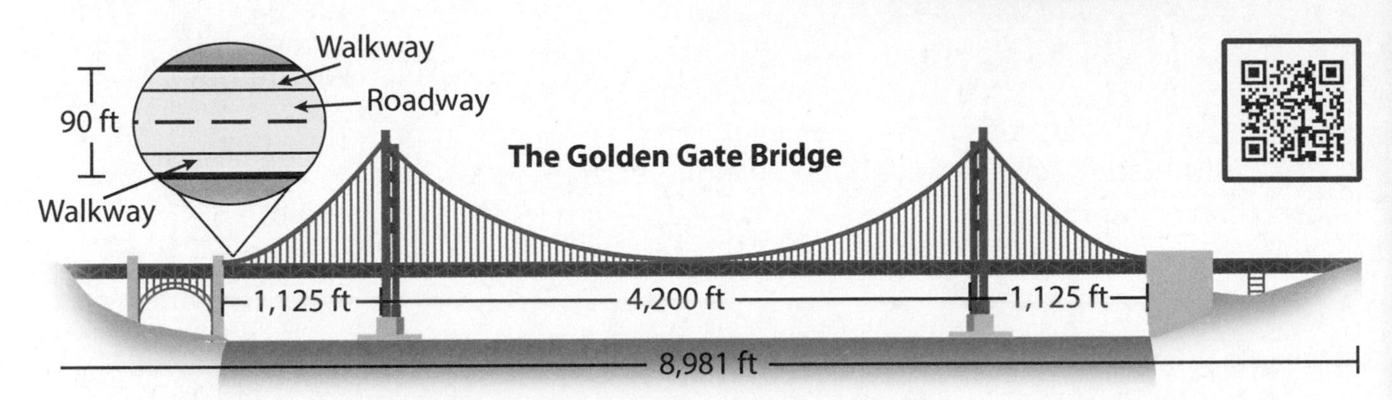

A teacher visits the Golden Gate Bridge in San Francisco, California.

1. Use the figure above to answer the questions.

a. The roadway is 8 times as wide as width of both walkways combined. Each walkway has the same width. How wide is one of the walkways?

b. What is the length of the bridge that is suspended above the water?

c. What is the length of the bridge that is *not* suspended above the water?

---

2. The teacher wants to estimate the distance between the bridge and the water.

a. He rides in a ferry boat under the bridge. He says that the distance between the bridge and the water is about 5 times the height of the ferry boat. What is the distance between the bridge and the water?

b. The teacher estimates that the height of one tower is about three times the distance between the bridge and the water. What is the teacher's estimate for the height of the tower?

c. You learn the exact height of the tower is 86 feet taller than the teacher's estimate. How tall is the tower?

The teacher uses his fitness tracker to count the number of steps he walks on the Golden Gate Bridge each day for 1 week.

Day	Number of Steps
Sunday	12,378
Monday	10,712
Tuesday	7,989
Wednesday	
Thursday	
Friday	
Saturday	

3. The table shows the numbers of steps taken on the first 3 days of the week. Use this information to complete the table.

a. He takes one thousand one hundred twenty-two more steps on Wednesday than on Monday.

b. On Thursday, he takes one hundred eighteen steps less than on Sunday.

c. He takes ten thousand fifty steps on Friday.

d. On Saturday, he takes twice as many steps as on Tuesday.

4. The teacher's goal is to take 11,500 steps each day.

a. What is his goal for the week?

b. Estimate the total number of steps he takes from Sunday to Saturday. Does he meet his goal for the week? Explain.

c. One mile is about two thousand steps. The teacher estimates that he walks 3 miles from one end of the Golden Gate Bridge to the other and back again. About how many steps does he walk?

4 Multiply by Two-Digit Numbers

- Wind turbines convert wind into energy. The wind is a renewable source of energy. What are some other forms of renewable energy?
- The amount of energy generated is based on the speed of the wind and the lengths of the turbine blades. What effect do you think the blade lengths have on the amount of energy a wind turbine can generate?

Chapter Learning Target:
Understand multiplying two-digit numbers.

Chapter Success Criteria:
- I can find the product of two numbers.
- I can use rounding to estimate a product.
- I can write multiplication problems.
- I can solve a problem using an equation.

Name ______________________________

4 Vocabulary

Review Words
place value
thousands

Organize It

Use the review words to complete the graphic organizer.

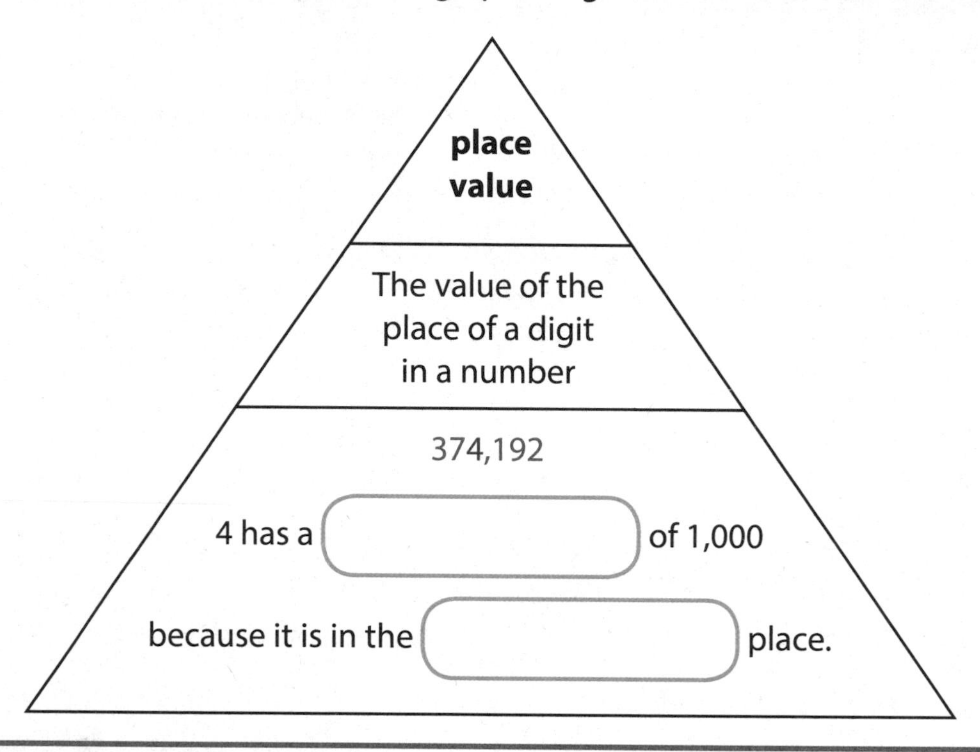

Define It

What am I?

Numbers that are easy to multiply and are close to the actual numbers

$3 \times 7 = T$	$6 \times 5 = A$	$5 \times 9 = C$	$3 \times 8 = M$	$4 \times 5 = O$
$9 \times 2 = B$	$8 \times 9 = N$	$7 \times 6 = P$	$8 \times 6 = I$	$7 \times 10 = L$
$7 \times 4 = R$	$4 \times 2 = E$	$10 \times 5 = U$	$6 \times 9 = S$	

45	20	24	42	30	21	48	18	70	8

72	50	24	18	8	28	54

Chapter 4 Vocabulary Cards

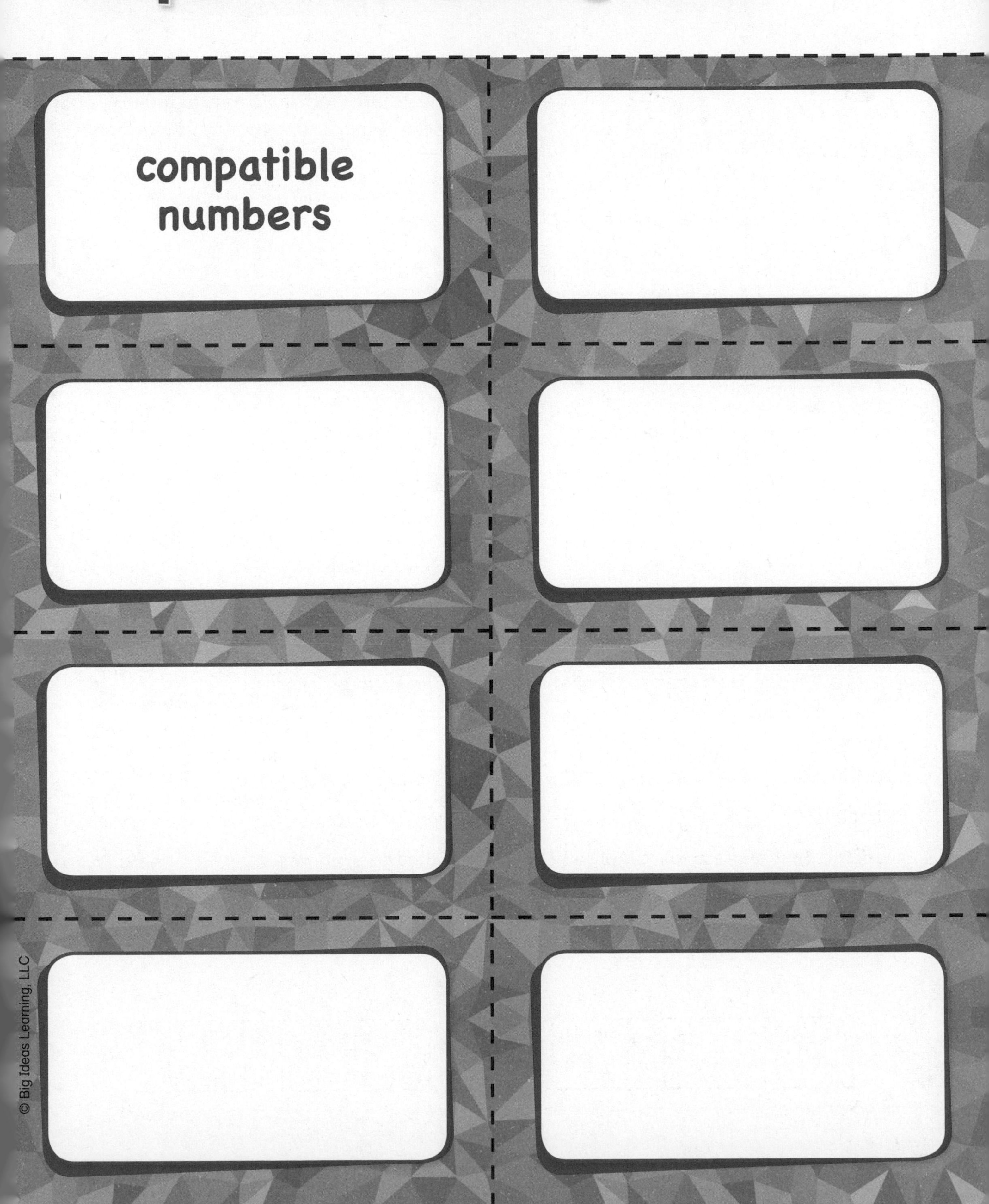

Numbers that are easy to multiply and are close to the actual numbers

24×31

↓ ↓

25×30

Name ______________________________

Learning Target: Use place value and properties to multiply by multiples of ten.

Success Criteria:

- I can use place value to multiply by multiples of ten.
- I can use the Associative Property to multiply by multiples of ten.
- I can describe a pattern with zeros when multiplying by multiples of ten.

Model each product. Draw each model.

$2 \times 3 =$ _____	$2 \times 30 =$ _____
$2 \times 300 =$ _____	$2 \times 3{,}000 =$ _____

What pattern do you notice in the products?

Repeated Reasoning How can the pattern above help you find 20×30?

Think and Grow: Multiply by Multiples of Tens

You can use place value and properties to multiply two-digit numbers by multiples of ten.

Example Find 40×20.

One Way: Use place value.

$40 \times 20 = 40 \times$ _____ tens

$=$ _____ tens

$=$ _____

So, $40 \times 20 =$ _____.

Another Way: Use the Associative Property of Multiplication.

$40 \times 20 = 40 \times (2 \times 10)$ Rewrite 20 as 2×10.

$= (40 \times 2) \times 10$ Associative Property of Multiplication

$=$ _____ $\times 10$

$=$ _____

So, $40 \times 20 =$ _____.

Example Find 12×30.

One Way: Use place value.

$12 \times 30 = 12 \times$ _____ tens

$=$ _____ tens

$=$ _____

So, $12 \times 30 =$ _____.

Another Way: Use the Associative Property of Multiplication.

$12 \times 30 = 12 \times (3 \times 10)$ Rewrite 30 as 3×10.

$= (12 \times 3) \times 10$ Associative Property of Multiplication

$=$ _____ $\times 10$

$=$ _____

So, $12 \times 30 =$ _____.

Show and Grow *I can do it!*

Find the product.

1. $70 \times 40 =$ _____

2. $50 \times 80 =$ _____

3. $24 \times 90 =$ _____

4. $45 \times 60 =$ _____

Name ______________________________

Apply and Grow: Practice

Find the product.

5. $90 \times 10 =$ _____	**6.** $40 \times 60 =$ _____	**7.** $20 \times 70 =$ _____
8. $11 \times 30 =$ _____	**9.** $12 \times 40 =$ _____	**10.** $15 \times 50 =$ _____
11. $30 \times 13 =$ _____	**12.** $10 \times 76 =$ _____	**13.** $40 \times 25 =$ _____

Find the missing factor.

14. $50 \times$ _____ $= 1{,}500$	**15.** $20 \times$ _____ $= 1{,}800$	**16.** $60 \times$ _____ $= 4{,}200$
17. _____ $\times 80 = 6{,}400$	**18.** _____ $\times 90 = 3{,}600$	**19.** _____ $\times 70 = 3{,}500$

Compare.

20. 60×30 ◯ $1{,}800$	**21.** 40×12 ◯ 460	**22.** 25×90 ◯ $2{,}340$

23. It takes 10 days to film 1 episode of a television show. How many days will it take to film a 20-episode season?

24. MP **Reasoning** What is Descartes's number? Explain.

The product of my number and 60 is 3,000.

25. **YOU BE THE TEACHER** Newton says that the product of two multiples of ten will always have exactly two zeros. Is he correct? Explain.

Think and Grow: Modeling Real Life

Example Food drive volunteers collect 1,328 cans of food. The volunteers have 50 boxes. Each box holds 20 cans. How many cans will *not* fit in the boxes?

Multiply to find how many cans will fit in the boxes.

$20 \times 50 = 20 \times (5 \times 10)$ Rewrite 50 as 5×10.

$= (20 \times 5) \times 10$ Associative Property of Multiplication

$=$ _____ $\times 10$

$=$ _____

So, _____ cans fit in the boxes.

Subtract the number of cans that will fit in the boxes from the total number of cans collected.

1,328
− ☐

_____ cans will not fit in the boxes.

Show and Grow *I can think deeper!*

26. A library has 2,124 new books. The library has 40 empty shelves. Each shelf holds 35 books. How many books will *not* fit on the empty shelves?

27. An apartment building has 15 floors. Each floor is 10 feet tall. An office building has 30 floors. Each floor is 13 feet tall. How much taller is the office building than the apartment building?

28. You burn 35 calories each hour you spend reading and 50 calories each hour you spend playing board games. In 2 weeks, you spend 14 hours reading and 28 hours playing board games. How many calories do you burn reading and playing board games?

Name ______________________________

Homework & Practice 4.1

Learning Target: Use place value and properties to multiply by multiples of ten.

Example Find 60×15.

One Way: Use place value.

$60 \times 15 =$ 6 tens $\times 15$

$=$ 90 tens

$=$ 900

So, $60 \times 15 =$ 900.

Another Way: Use the Associative Property of Multiplication.

$60 \times 15 = (10 \times 6) \times 15$ Rewrite 60 as 10×6.

$= 10 \times (6 \times 15)$ Associative Property of Multiplication

$= 10 \times$ 90

$=$ 900

So, $60 \times 15 =$ 900.

Find the product.

1. $30 \times 10 =$ ______
2. $20 \times 90 =$ ______
3. $50 \times 70 =$ ______
4. $40 \times 13 =$ ______
5. $27 \times 60 =$ ______
6. $80 \times 56 =$ ______

Find the missing factor.

7. $70 \times$ ______ $= 2{,}100$
8. ______ $\times 10 = 900$
9. $40 \times$ ______ $= 1{,}600$
10. ______ $\times 20 = 1{,}600$
11. $30 \times$ ______ $= 1{,}800$
12. ______ $\times 50 = 3{,}000$

Compare.

13. 90×80 ◯ $8{,}100$
14. $1{,}200$ ◯ 60×17
15. 34×70 ◯ $2{,}380$

16. A shallow moonquake occurs 20 kilometers below the moon's surface. A deep moonquake occurs 35 times deeper than a shallow moonquake. How many kilometers below the surface does the deep moonquake occur?

17. **MP Structure** Write the multiplication equation represented by the number line.

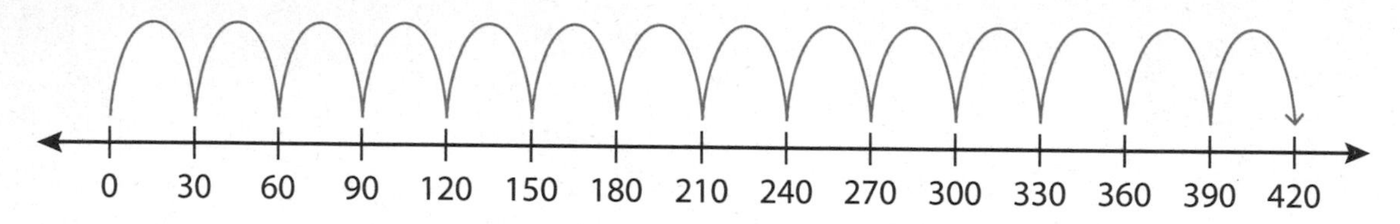

$\_\_\_\_ \times \_\_\_\_ = \_\_\_\_$

18. **Writing** Explain how you can use $20 \times 10 = 200$ to find 20×12.

19. **DIG DEEPER!** The product of a number and twice that number is 800. What are the numbers?

20. **Modeling Real Life** There are 506 new plants in a greenhouse. A worker programs a robot to arrange the plants into 14 rows with 30 plants in each row. How many plants will *not* fit in the rows?

21. **Modeling Real Life** The world's largest pool is 13 meters longer than the total length of 20 Olympic pools. An Olympic pool is 50 meters long. How long is the world's largest pool?

Review & Refresh

Find the value of the underlined digit.

22. 52,61$\underline{8}$

23. 3$\underline{7}$9,021

24. $\underline{2}$03,557

25. 497,$\underline{3}$84

Name ______________________________

Estimate Products 4.2

Learning Target: Use rounding and compatible numbers to estimate products.

Success Criteria:

- I can use rounding to estimate a product.
- I can use compatible numbers to estimate a product.
- I can explain different ways to estimate a product.

Explore and Grow

Choose an expression to estimate each product. Write the expression. You may use an expression more than once.

20×20 $\quad$ 20×25

25×40 $\quad$ 40×20

21×24	26×38	23×17	42×23
____ × ____	____ × ____	____ × ____	____ × ____

Compare your answers with a partner. Did you choose the same expressions?

Construct Arguments Which estimated product do you think will be closer to the product of 29 and 37? Explain your reasoning.

25×40

30×40

Think and Grow: Estimate Products

You can estimate products using rounding or compatible numbers.

Compatible numbers are numbers that are easy to multiply and are close to the actual numbers.

Example Use rounding to estimate 57×38.

Step 1: Round each factor to the nearest ten.

57×38

↓ ↓

60×40

Step 2: Multiply.

$60 \times 40 = 60 \times$ ______ tens

$=$ ______ tens

$=$ ______

So, 57×38 is about ______.

Example Use compatible numbers to estimate 24×31.

Step 1: Choose compatible numbers.

Think: 24 is close to 25.
31 is close to 30.

24×31

↓ ↓

25×30

Step 2: Multiply.

$25 \times 30 = 25 \times$ ______ tens

$=$ ______ tens

$=$ ______

So, 24×31 is about ______.

Show and Grow *I can do it!*

Use rounding to estimate the product.

1. 27×50

2. 42×14

3. 61×73

Use compatible numbers to estimate the product.

4. 19×26

5. 23×78

6. 74×20

Name ______________________________

Apply and Grow: Practice

Estimate the product.

7. 41×73

8. 52×84

9. 26×68

10. 38×17

11. 75×24

12. 93×53

13. 44×78

14. 21×33

15. 45×45

Open-Ended Write two possible factors that can be estimated as shown.

16. 2,400

_____ × _____

↓ ↓

_____ × _____ = 2,400

17. 1,200

_____ × _____

↓ ↓

_____ × _____ = 1,200

18. **DIG DEEPER!** You use 50×30 to estimate 46×29. Will your estimate be greater than or less than the actual product? Explain.

19. **YOU BE THE TEACHER** Your friend uses rounding to estimate 15×72. She gets a product of 700. Is your friend's estimate correct? Explain.

Think and Grow: Modeling Real Life

Example About how much does 1 year of phone service cost?

Service Provider Rates	
Service	**Price per Month**
Phone	\$24
Internet	\$45
Cable television	\$89

Think: What do you know? What do you need to find? How will you solve?

There are 12 months in 1 year, so multiply the price per month by 12.

$$12 \times 24 = ?$$

Estimate the product.

One Way: Use rounding to estimate.

12 × \$24
↓ ↓
10 × \$20 = \$_____

So, 1 year of phone service costs about \$_____.

Another Way: Use compatible numbers to estimate.

12 × \$24
↓ ↓
10 × \$25 = \$_____

So, 1 year of phone service costs about \$_____.

Show and Grow *I can think deeper!*

20. Use the table above. About how much does 1 year of Internet service cost? About how much does 1 year of cable television service cost?

21. A giant panda eats 28 pounds of food each day. An orca eats 17 times as much food as the panda eats each day. About how much food does the orca eat each day?

Name ______________________________

Homework & Practice 4.2

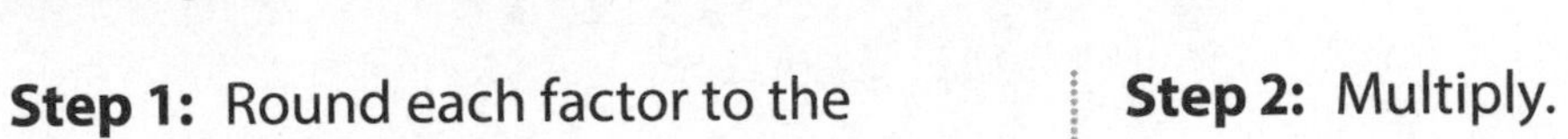

Learning Target: Use rounding and compatible numbers to estimate products.

Example Use rounding to estimate 45×43.

Step 1: Round each factor to the nearest ten.

45×43

↓ ↓

50×40

Step 2: Multiply.

$50 \times 40 = 50 \times$ 4 tens

= 200 tens

= 2,000

So, 45×43 is about 2,000.

Example Use compatible numbers to estimate 61×24.

Step 1: Choose compatible numbers.

Think: 61 is close to 60.
24 is close to 25.

61×24

↓ ↓

60×25

Step 2: Multiply.

$60 \times 25 =$ 6 tens $\times 25$

= 150 tens

= 1,500

So, 61×24 is about 1,500.

Use rounding to estimate the product.

1. 42×13

2. 56×59

3. 19×91

Use compatible numbers to estimate the product.

4. 23×78

5. 67×45

6. 19×24

Estimate the product.

7. 84×78

8. 92×34

9. 57×81

Open-Ended Write two possible factors that could be estimated as shown.

10. 6,400

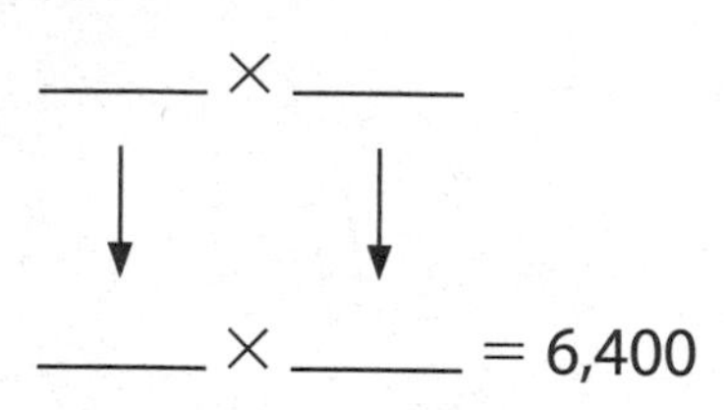

11. 1,600

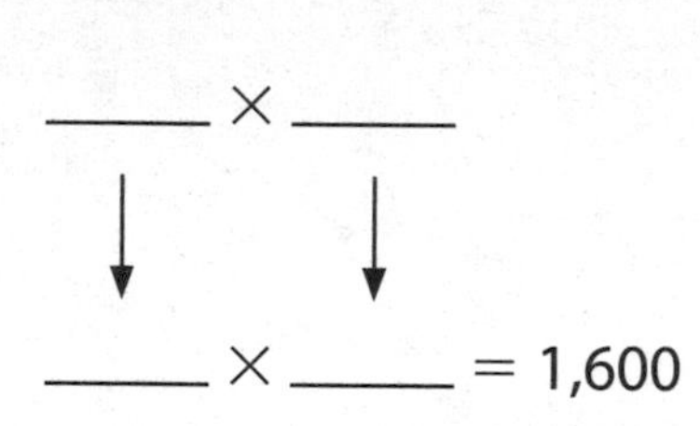

12. MP **Reasoning** Are both Newton's and Descartes's estimates reasonable? Explain.

13. **DIG DEEPER!** You use 90×30 to estimate 92×34. Will your estimate be greater than or less than the actual product? Explain.

14. **Modeling Real Life** About how many hours of darkness does Barrow, Alaska have in December?

Days of Darkness in Barrow, Alaska

Month	Days
November	12
December	31
January	21

Review & Refresh

15. Round 253,490 to the nearest ten thousand.

16. Round 628,496 to the nearest hundred thousand.

Name ___________________________________

Use Area Models to Multiply Two-Digit Numbers 4.3

Learning Target: Use area models and partial products to multiply.

Success Criteria:

- I can use an area model to break apart the factors of a product.
- I can relate an area model to partial products.
- I can add partial products to find a product.

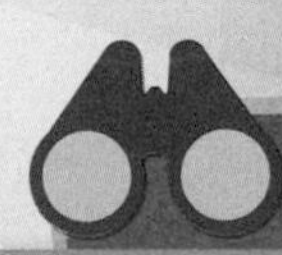

Explore and Grow

Draw an area model that represents 15×18. Then break apart your model into smaller rectangles.

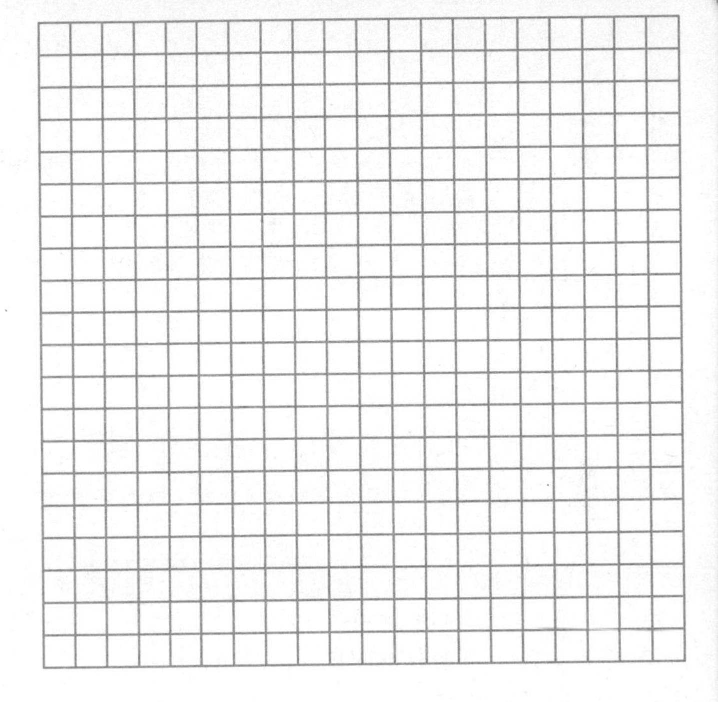

What is the total area of your model? Explain how you found your answer.

MP Reasoning Compare with a partner. Do you get the same answer? Explain.

Think and Grow: Use Area Models to Multiply

Why does the sum of the partial products represent the sum of the whole area?

Example Use an area model and partial products to find 12 × 14.

Model the expression. Break apart 12 as 10 + 2 and 14 as 10 + 4.

Add the area of each rectangle to find the product for the whole model.

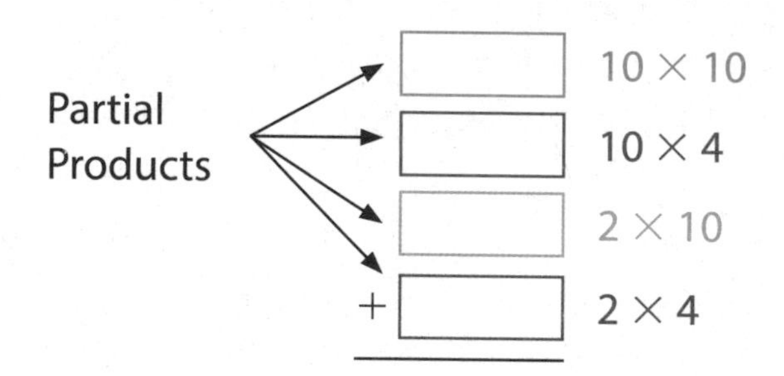

_____ Add the partial products.

So, 12 × 14 = _____.

Show and Grow *I can do it!*

Use the area model to find the product.

1. 17 × 15 = _____

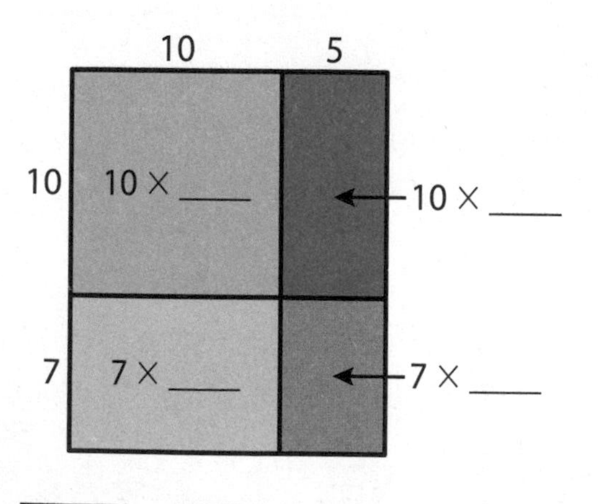

☐ + ☐ + ☐ + ☐

2. 34 × 22 = _____

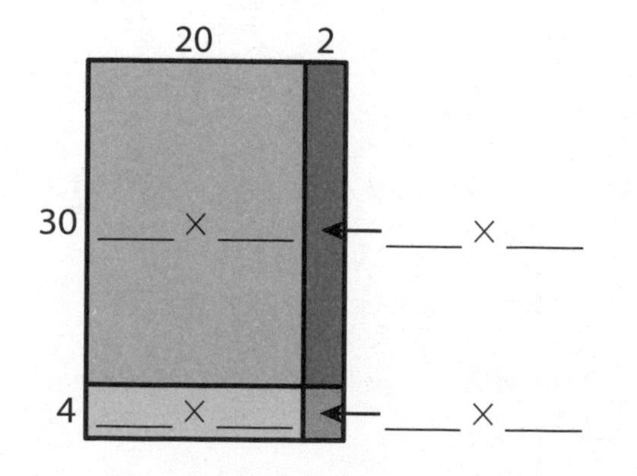

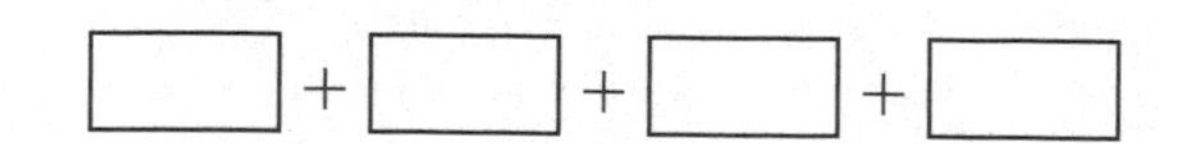

Name ______________________________

Apply and Grow: Practice

Use the area model to find the product.

3. $13 \times 19 =$ ______

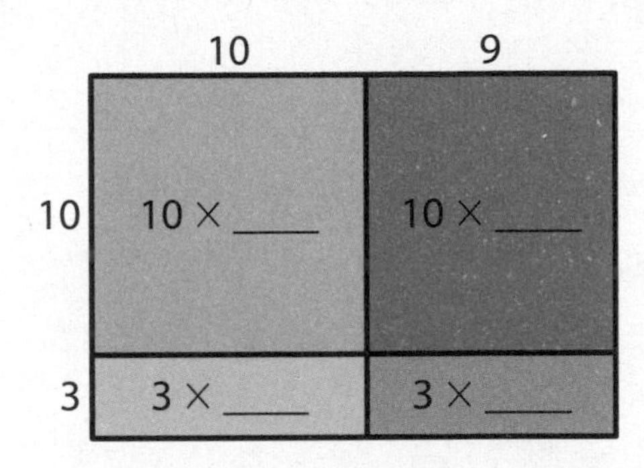

4. $25 \times 39 =$ ______

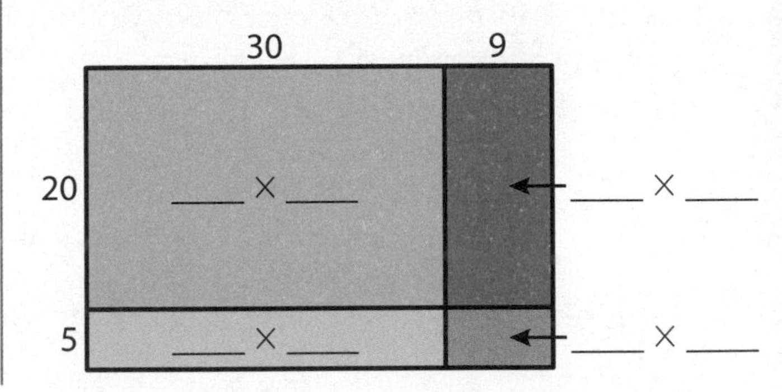

Draw an area model to find the product.

5. $11 \times 13 =$ ______

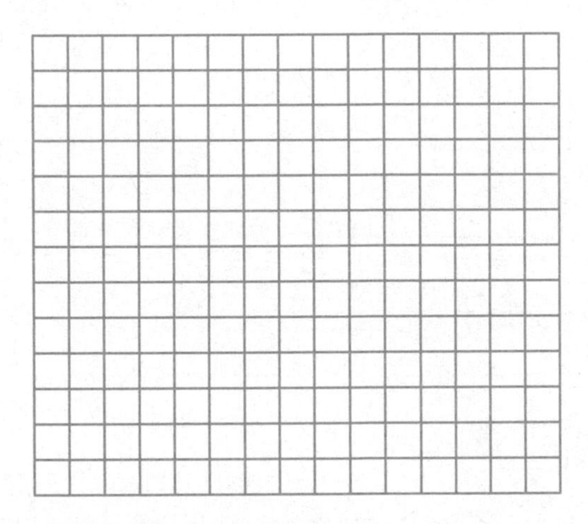

6. $23 \times 26 =$ ______

7. $27 \times 45 =$ ______

8. Perseid meteors travel 59 kilometers each second. How far does a perseid meteor travel in 15 seconds?

9. **DIG DEEPER!** Write the multiplication equation represented by the area model.

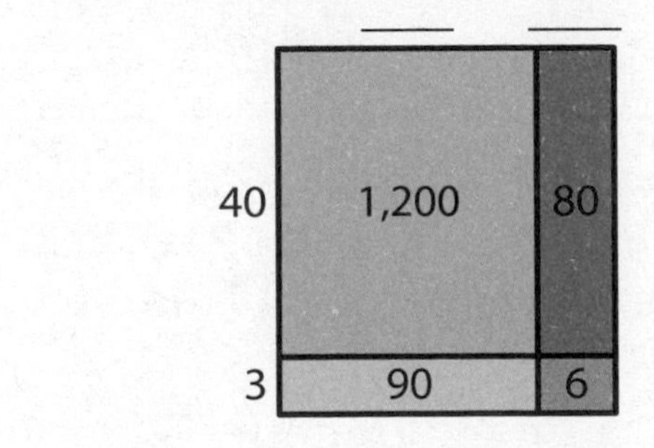

____ × ____ = ____

Think and Grow: Modeling Real Life

Example A wind farm has 8 rows of new wind turbines and 3 rows of old wind turbines. Each row has 16 turbines. How many turbines does the wind farm have?

Add the number of rows of new turbines to the number of rows of old turbines.

$8 + 3 =$ ______

There are ______ rows of turbines.

Multiply the number of rows by the number in each row.

11×16

	10	6
10	10×10	10×6
1	1×10	1×6

☐	10×10
☐	10×6
☐	1×10
+ ☐	1×6

______ Add the partial products.

The wind farm has ______ turbines.

Show and Grow *I can think deeper!*

10. You can type 19 words per minute. Your cousin can type 33 words per minute. How many more words can your cousin type in 15 minutes than you?

11. A store owner buys 24 packs of solar eclipse glasses. Each pack has 12 glasses. The store did *not* sell 18 of the glasses. How many of the glasses did the store sell?

Name ______________________________

Homework & Practice 4.3

Learning Target: Use area models and partial products to multiply.

Example Use an area model and partial products to find 15 × 18.

Model the expression. Break apart 15 as 10 + 5 and 18 as 10 + 8.

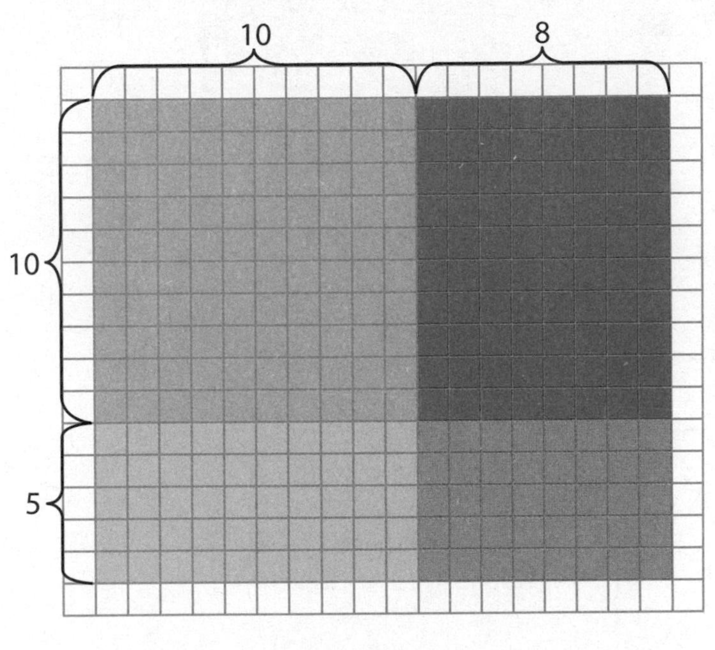

Add the area of each rectangle to find the product for the whole model.

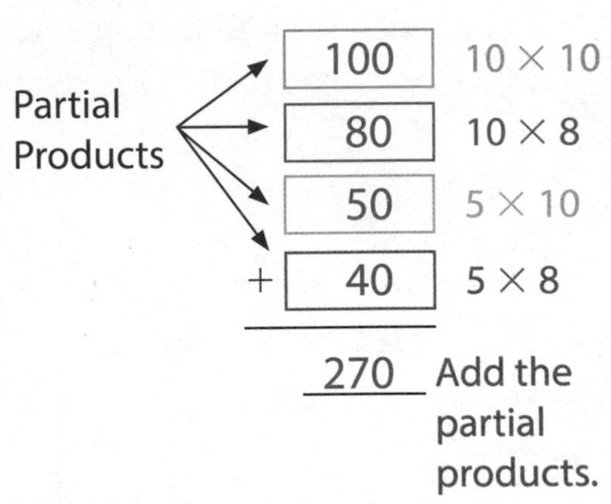

So, 15 × 18 = 270.

Use the area model to find the product.

1. 12 × 13 = ______

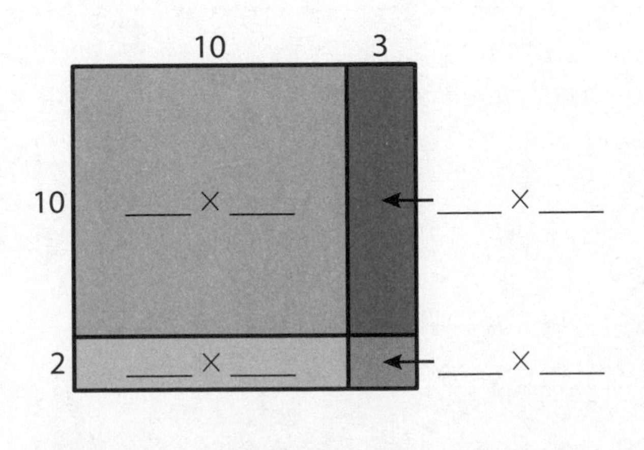

☐ + ☐ + ☐ + ☐

2. 38 × 24 = ______

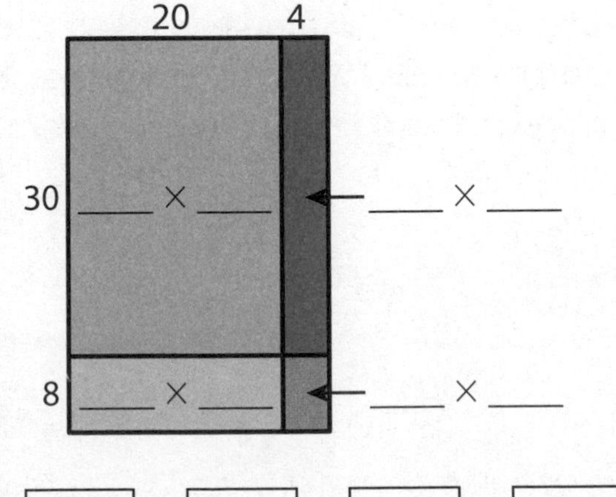

☐ + ☐ + ☐ + ☐

Use the area model to find the product.

3. $19 \times 18 =$ _____

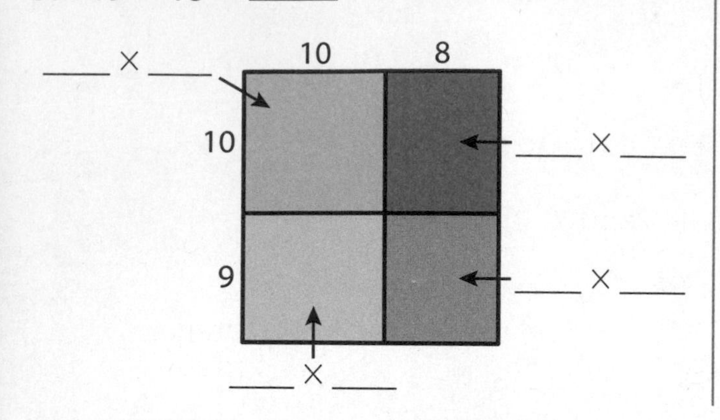

4. $23 \times 25 =$ _____

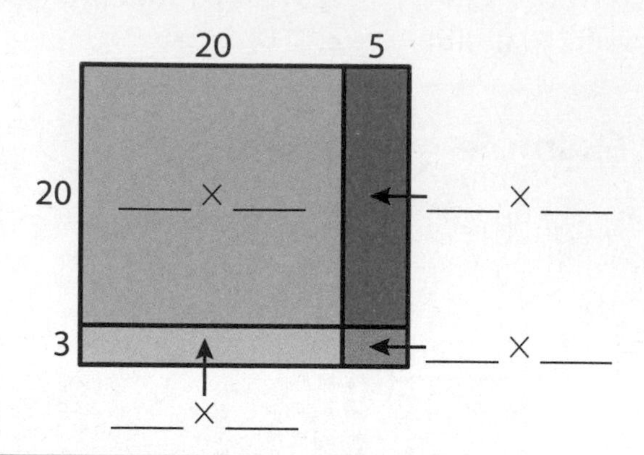

Draw an area model to find the product.

5. $26 \times 31 =$ _____

6. $22 \times 47 =$ _____

7. **YOU BE THE TEACHER** Your friend finds 12×42. Is your friend correct? Explain.

	40	2
10	400	20
2	80	4

$400 + 80 + 20 + 4 = 504$

8. **Writing** Explain how to use an area model and partial products to multiply two-digit numbers.

9. **Modeling Real Life** A mega-arcade has 9 rows of single-player games and 5 rows of multi-player games. Each row has 24 games. How many games does the arcade have?

Review & Refresh

Find the sum. Check whether your answer is reasonable.

10. $75{,}420 + 8{,}596$

11. $47{,}928 + 23{,}657$

12. $505{,}019 + 64{,}802$

Name ____________________

Use the Distributive Property to Multiply Two-Digit Numbers

4.4

Learning Target: Use area models and the Distributive Property to multiply.

Success Criteria:

- I can use an area model and partial products to multiply.
- I can use an area model and the Distributive Property to multiply.

Explore and Grow

Use as few base ten blocks as possible to create an area model for 13×24. Draw to show your model.

24

13

Color your model to show four smaller rectangles. Label the partial products.

Reasoning How do you think the Distributive Property relates to your area model? Explain.

Think and Grow: Use the Distributive Property to Multiply

Example Find 17×25.

One Way: Use an area model and partial products.

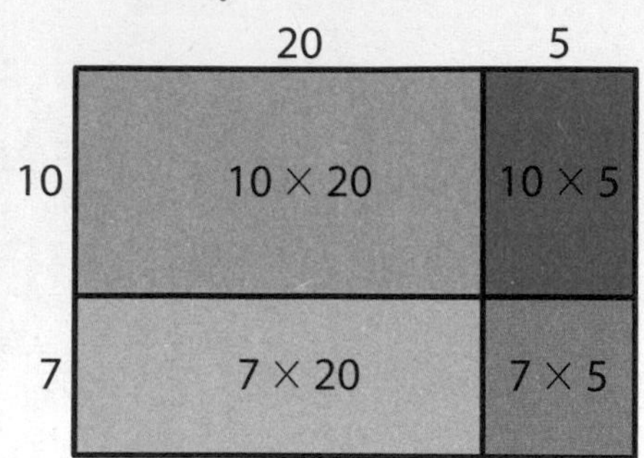

Add the partial products.

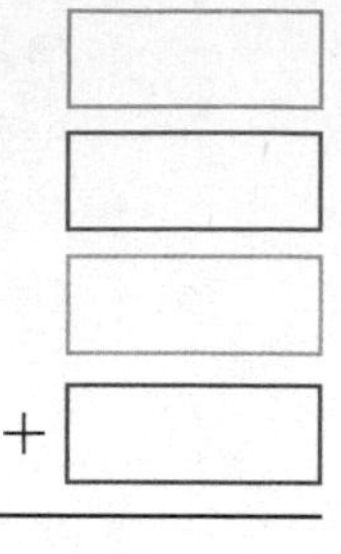

So, $17 \times 25 =$ ______.

Another Way: Use an area model and the Distributive Property.

$17 \times 25 = 17 \times (20 + 5)$ Break apart 25.

$= (17 \times 20) + (17 \times 5)$ Distributive Property

$= (10 + 7) \times 20 + (10 + 7) \times 5$ Break apart 17.

$= (10 \times 20) + (7 \times 20) + (10 \times 5) + (7 \times 5)$ Distributive Property

$=$ ______ $+$ ______ $+$ ______ $+$ ______

$=$ ______

So, $17 \times 25 =$ ______.

Show and Grow *I can do it!*

1. Use the area model and the Distributive Property to find 32×19.

	10	9
30	30 × 10	30 × 9
2	2 × 10	2 × 9

$32 \times 19 = 32 \times (10 + 9)$

$= (32 \times 10) + (32 \times 9)$

$= (30 + 2) \times$ ______ $+ (30 + 2) \times$ ______

$= (30 \times$ ______$) + (2 \times$ ______$) + (30 \times$ ______$) + (2 \times$ ______$)$

$=$ ______ $+$ ______ $+$ ______ $+$ ______

$=$ ______

So, $32 \times 19 =$ ______.

Name ______________________________

Apply and Grow: Practice

2. Use the area model and the Distributive Property to find 34×26.

	20	6
30	30×20	30×6
4	4×20	4×6

$34 \times 26 = 34 \times (20 + 6)$

$= (34 \times 20) + (34 \times 6)$

$= (30 + 4) \times$ _____ $+ (30 + 4) \times$ _____

$= (30 \times$ _____$) + (4 \times$ _____$) + (30 \times$ _____$) + (4 \times$ _____$)$

$=$ _____ $+$ _____ $+$ _____ $+$ _____

$=$ _____

So, $34 \times 26 =$ _____.

Use the Distributive Property to find the product.

3. $28 \times 47 = 28 \times (40 + 7)$

$= (28 \times 40) + (28 \times 7)$

$= (20 + 8) \times$ _____ $+ (20 + 8) \times$ _____

$= (20 \times$ _____$) + (8 \times$ _____$) + (20 \times$ _____$) + (8 \times$ _____$)$

$=$ _____ $+$ _____ $+$ _____ $+$ _____

$=$ _____

So, $28 \times 47 =$ _____.

4. $39 \times 41 =$ _____

5. $74 \times 12 =$ _____

6. $83 \times 65 =$ _____

7. Which One Doesn't Belong? Which expression does *not* belong with the other three?

$(40 + 7) \times 52$ $\quad$ $(40 + 7) \times (50 + 2)$ $\quad$ $(40 \times 7) \times (50 \times 2)$ $\quad$ $47 \times (50 + 2)$

Think and Grow: Modeling Real Life

Example The dunk tank at a school fair needs 350 gallons of water. There are 27 students in a class. Each student pours 13 gallons of water into the tank. Is there enough water in the dunk tank?

Find how many gallons of water the students put in the dunk tank.

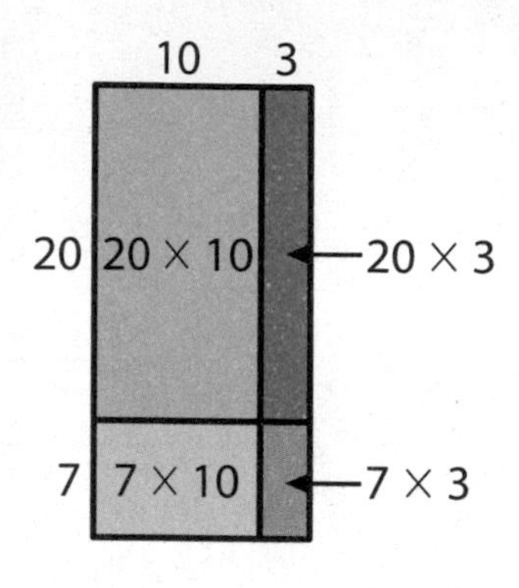

$27 \times 13 = 27 \times (10 + 3)$

$= (27 \times 10) + (27 \times 3)$

$= (20 + 7) \times 10 + (20 + 7) \times 3$

$= (20 \times 10) + (7 \times 10) + (20 \times 3) + (7 \times 3)$

= _____ + _____ + _____ + _____

= _____ gallons

Compare the numbers of gallons.

So, there _____ enough water in the dunk tank.

Show and Grow *I can think deeper!*

8. An event coordinator orders 35 boxes of T-shirts to give away at a baseball game. There are 48 T-shirts in each box. If 2,134 fans attend the game, will each fan get a T-shirt?

9. A horse owner must provide 4,046 square meters of pasture for each horse. Is the pasture large enough for 2 horses? Explain.

Name ______________________________

Homework & Practice 4.4

Learning Target: Use area models and the Distributive Property to multiply.

Example Find 13×28.

One Way: Use an area model and partial products.

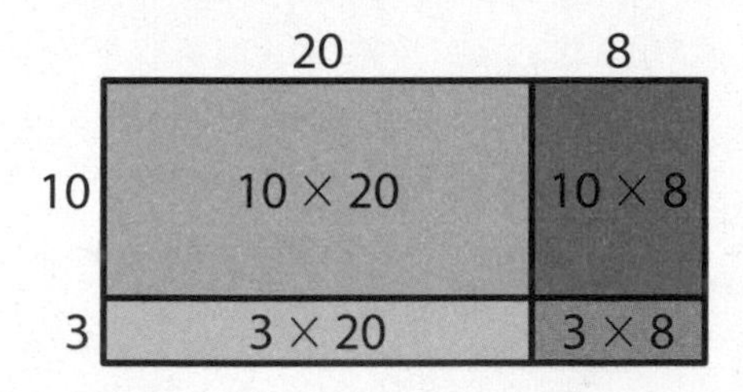

Add the partial products.

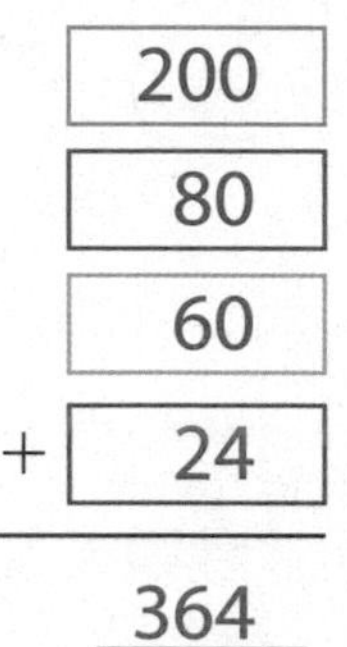

So, $13 \times 28 =$ 364.

Another Way: Use an area model and the Distributive Property.

$13 \times 28 = 13 \times (20 + 8)$	Break apart 28.
$= (13 \times 20) + (13 \times 8)$	Distributive Property
$= (10 + 3) \times 20 + (10 + 3) \times 8$	Break apart 13.
$= (10 \times 20) + (3 \times 20) + (10 \times 8) + (3 \times 8)$	Distributive Property
$=$ 200 + 60 + 80 + 24	
$=$ 364	So, $13 \times 28 =$ 364.

1. Use the area model and the Distributive Property to find 45×21.

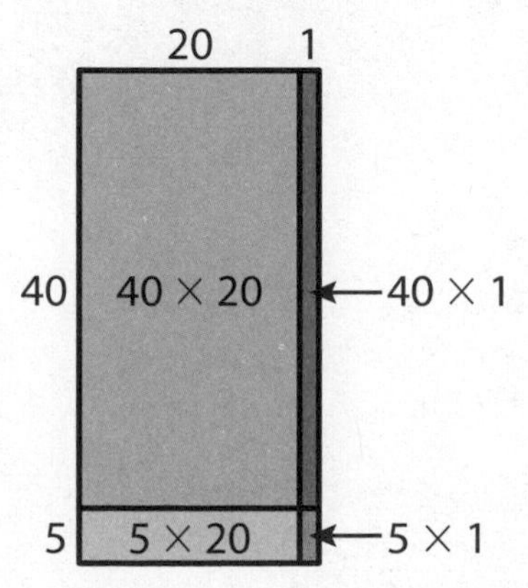

$45 \times 21 = 45 \times (20 + 1)$

$= (45 \times 20) + (45 \times 1)$

$= (40 + 5) \times$ ____ $+ (40 + 5) \times$ ____

$= (40 \times$ ____$) + (5 \times$ ____$) + (40 \times$ ____$) + (5 \times$ ____$)$

$=$ ____ $+$ ____ $+$ ____ $+$ ____

$=$ ____

So, $45 \times 21 =$ ____.

2. Use the Distributive Property to find the product.

$34 \times 49 = 34 \times (40 + 9)$

$= (34 \times 40) + (34 \times 9)$

$= (30 + 4) \times$ _____ $+ (30 + 4) \times$ _____

$= (30 \times$ _____$) + (4 \times$ _____$) + (30 \times$ _____$) + (4 \times$ _____$)$

$=$ _____ $+$ _____ $+$ _____ $+$ _____

$=$ _____

So, $34 \times 49 =$ _____.

3. $14 \times 27 =$ _____

4. $38 \times 31 =$ _____

5. $58 \times 26 =$ _____

6. $56 \times 32 =$ _____

7. $87 \times 23 =$ _____

8. $95 \times 81 =$ _____

9. **DIG DEEPER!** Find 42×78 by breaking apart 42 first.

10. **Modeling Real Life** The Elephant Building is 335 feet high. A real Asian elephant is 12 feet tall. If 29 real elephants could stand on top of each other, would they reach the top of the building?

Elephant Building
Bangkok, Thailand

Review & Refresh

Find the difference. Then check your answer.

11. $30{,}698 - 5{,}439$

12. $90{,}800 - 37{,}638$

13. $214{,}507 - 73{,}569$

Name ______________________________

Use Partial Products to Multiply Two-Digit Numbers 4.5

Learning Target: Use place value and partial products to multiply.

Success Criteria:

- I can use place value to tell the value of each digit in a number.
- I can write the partial products for a multiplication problem.
- I can add the partial products to find a product.

Explore and Grow

How can you use the rectangles to find 24×53? Complete the equation.

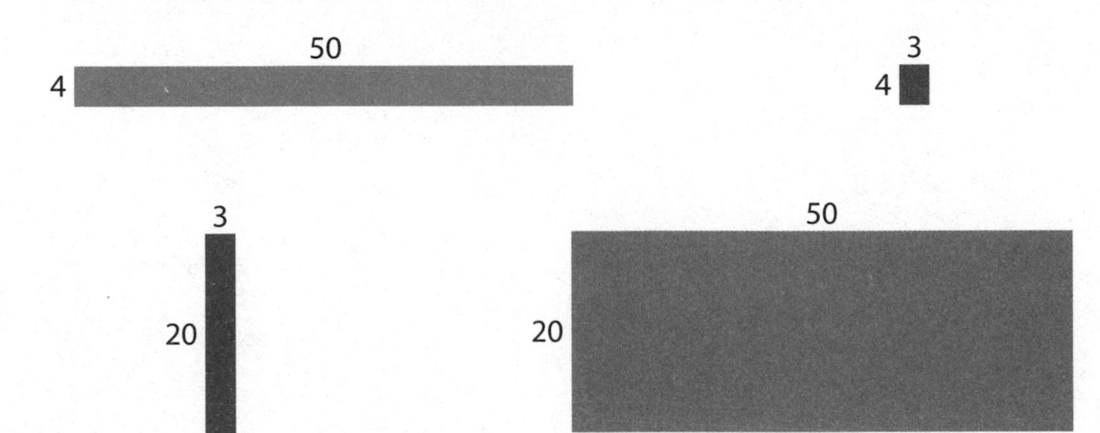

$24 \times 53 =$ _____

Reasoning What does the area of each rectangle represent?

Think and Grow: Use Partial Products to Multiply Two-Digit Numbers

Example Use place value and partial products to find 27 × 48.

Estimate: 30 × 50 = ______

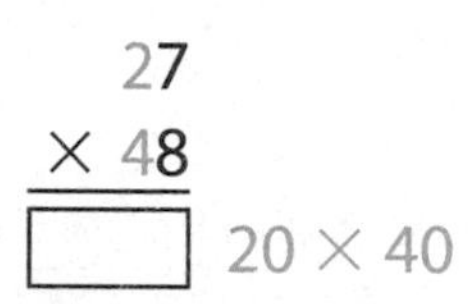

Step 1: Multiply the tens by the tens.

```
   27
 × 48
 [  ]   20 × 40
```

Step 2: Multiply the ones by the tens.

```
   27
 × 48
  800
 [  ]   7 × 40
```

Step 3: Multiply the tens by the ones.

```
   27
 × 48
  800
  280
 [  ]   20 × 8
```

Step 4: Multiply the ones by the ones.

```
   27
 × 48
  800
  280
  160
+ [  ]   7 × 8
  [  ]   Add the partial products.
```

So, 27 × 48 = ________.

Check: Because ________ is close to the estimate, ________, the answer is reasonable.

Show and Grow *I can do it!*

Find the product. Check whether your answer is reasonable.

1. Estimate: ________

```
   39
 × 15
 [  ]
 [  ]
 [  ]
+[  ]
```

2. Estimate: ________

```
   82
 × 63
 [  ]
 [  ]
 [  ]
+[  ]
```

3. Estimate: ________

```
   56
 × 71
 [  ]
 [  ]
 [  ]
+[  ]
```

Name ___

Apply and Grow: Practice

Find the product. Check whether your answer is reasonable.

4. Estimate: ___

$$\begin{array}{r} 14 \\ \times\ 26 \\ \hline \square \\ \square \\ \square \\ +\ \square \\ \hline \end{array}$$

5. Estimate: ___

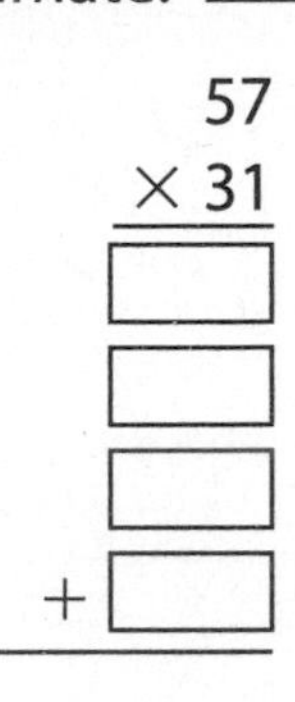

$$\begin{array}{r} 57 \\ \times\ 31 \\ \hline \square \\ \square \\ \square \\ +\ \square \\ \hline \end{array}$$

6. Estimate: ___

$$\begin{array}{r} 23 \\ \times\ 92 \\ \hline \square \\ \square \\ \square \\ +\ \square \\ \hline \end{array}$$

7. Estimate: ___

$13 \times 98 =$ ___

8. Estimate: ___

$65 \times 22 =$ ___

9. Estimate: ___

$72 \times 81 =$ ___

10. A farmer has 58 cows. Each cow produces 29 liters of milk. How many liters of milk do the cows produce in all?

11. MP **Number Sense** How much greater is the product of 12 and 82 than the product of 11 and 82? Explain how you know without multiplying.

12. DIG DEEPER! Write the multiplication equation shown by the partial products.

$200 + 60 + 50 + 15$

Think and Grow: Modeling Real Life

Daily Sleep Totals	
Koala	★ ★ ★ ★ ★ ½
Python	★ ★ ★ ★ ½
Tiger	★ ★ ★ ★

Each ★ = 4 hours.

Example How many hours does a koala sleep in 2 weeks?

Find how many hours a koala sleeps each day.

5 ★s = 5 × _____ = _____

1 ½★ = _____

_____ + _____ = _____

A koala sleeps _____ hours each day.

Multiply to find how many hours a koala sleeps in 2 weeks.

$$\begin{array}{r} 22 \\ \times\ 14 \\ \hline \end{array}$$

A koala sleeps _____ hours in 2 weeks.

Show and Grow *I can think deeper!*

13. Use the table above. How many hours does a python sleep in 3 weeks?

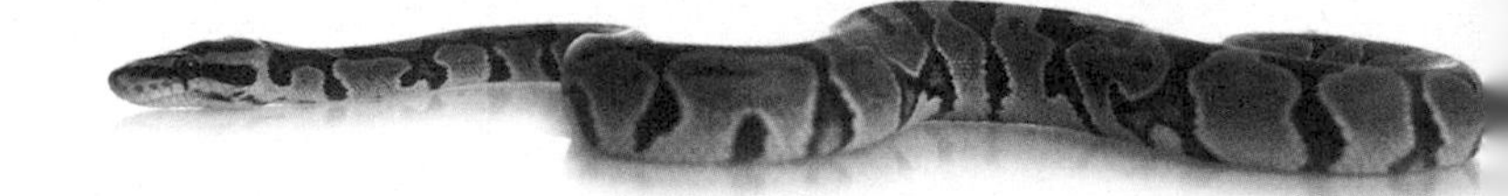

Seeds in Each Packet	
Cauliflower	⬮ ⬮ ⬮ ⬮ ⬮
Pumpkin	⬮ ⬮ ◖
Cucumber	⬮ ⬮ ⬮ ◖
Pea	⬮ ⬮ ⬮ ⬮ ⬮ ⬮

Each ⬮ = 12 seeds.

14. You have 12 packets of pea seeds and 23 packets of cucumber seeds. How many fewer pea seeds do you have than cucumber seeds?

Name ______________________________

Homework & Practice 4.5

Learning Target: Use place value and partial products to multiply.

Example Use place value and partial products to find 29 × 85.

Estimate: 30 × 85 = <u>2,550</u>

Step 1: Multiply the ones by the ones.

```
   29
 × 85
 [45]   9 × 5
```

Step 2: Multiply the tens by the ones.

```
   29
 × 85
   45
 [100]  20 × 5
```

Step 3: Multiply the ones by the tens.

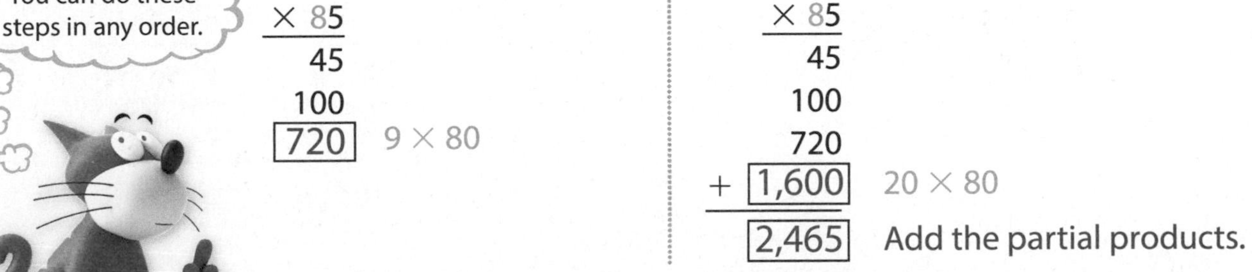


```
   29
 × 85
   45
  100
 [720]  9 × 80
```

Step 4: Multiply the tens by the tens.

```
     29
   × 85
     45
    100
    720
+ [1,600]  20 × 80
  [2,465]  Add the partial products.
```

So, 29 × 85 = <u>2,465</u>.

Check: Because <u>2,465</u> is close to the estimate, <u>2,550</u>, the answer is reasonable.

Find the product. Check whether your answer is reasonable.

1. Estimate: ________

```
   26
 × 17
 [  ]
 [  ]
 [  ]
+[  ]
```

2. Estimate: ________

```
   38
 × 62
 [  ]
 [  ]
 [  ]
+[  ]
```

3. Estimate: ________

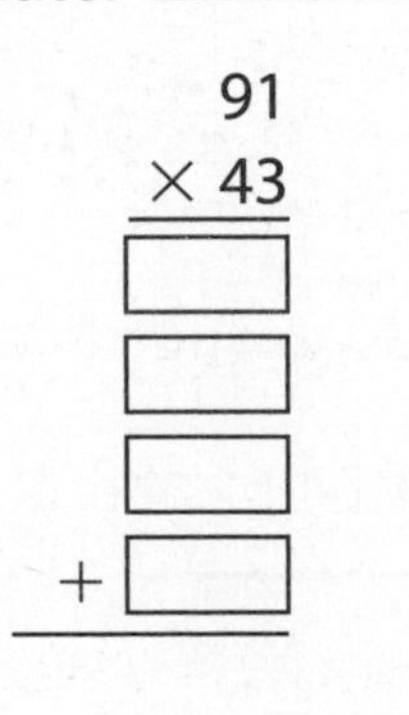

```
   91
 × 43
 [  ]
 [  ]
 [  ]
+[  ]
```

Find the product. Check whether your answer is reasonable.

4. Estimate: ________

$$\begin{array}{r} 51 \\ \times\ 74 \\ \hline \end{array}$$

5. Estimate: ________

$$\begin{array}{r} 28 \\ \times\ 19 \\ \hline \end{array}$$

6. Estimate: ________

$$\begin{array}{r} 35 \\ \times\ 45 \\ \hline \end{array}$$

7. Estimate: ________

$82 \times 63 =$ ________

8. Estimate: ________

$36 \times 93 =$ ________

9. Estimate: ________

$57 \times 22 =$ ________

10. **DIG DEEPER!** Find the missing digits. Then find the product.

$$\begin{array}{r} \square\,5 \\ \times \quad \square\,6 \\ \hline 1\,0\,0 \\ 5\,0 \\ 6\,0 \\ +\quad 3\,0 \\ \hline \boxed{} \end{array}$$

11. **Modeling Real Life** If Newton meets his goal each month, how many liters of water will he drink in 1 year?

Monthly Water Intake Goals	
You	★ ★ ★ ★ ★ ★ ★ ½
Newton	★ ★ ★ ★ ★ ½
Descartes	★

Each ★ = 6 liters.

12. **Modeling Real Life** Use the table in Exercise 11. If you and Descartes each meet your goal each month, how many more liters will you drink in 1 year than Descartes?

Review & Refresh

Add or subtract. Then check your answer.

13. $512{,}006 + 318{,}071 =$ ________

14. $746{,}620 - 529{,}706 =$ ________

Name ________________________________

Multiply Two-Digit Numbers 4.6

Learning Target: Multiply two-digit numbers.

Success Criteria:

- I can multiply to find partial products.
- I can show how to regroup ones, tens, and hundreds.
- I can add partial products to find a product.

Use base ten blocks to model each product. Draw each model.

$10 \times 10 =$ ______	$10 \times 3 =$ ______
$7 \times 10 =$ ______	$7 \times 3 =$ ______

Reasoning How are the models related to the product 17×13?

Think and Grow: Use Regrouping to Multiply Two-Digit Numbers

Example Find 87 × 64.

Estimate: 90 × 60 = ________

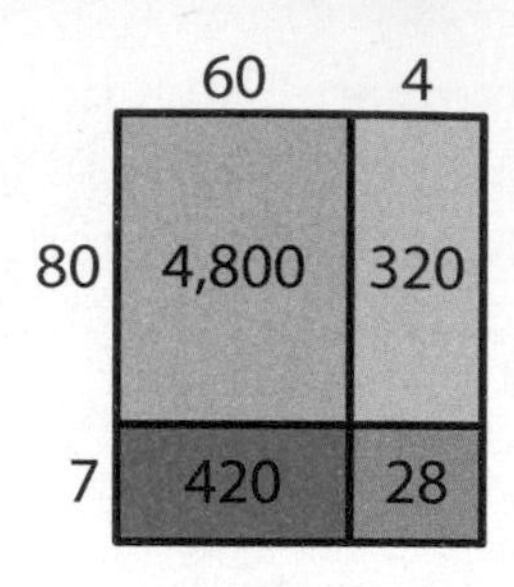

Step 1: Multiply 87 by 4 ones, or 4.

```
       2
      87
   ×  64
4 × 87 → 348
```

- 4 × 7 ones = 28 ones
  Regroup as 2 tens and 8 ones.
- 4 × 8 tens = 32 tens
  Add the regrouped tens: 32 + 2 = 34.

Step 2: Multiply 87 by 6 tens, or 60.

```
        4
        2
       87
   ×   64
      348
60 × 87 → 5,220
```

- 6 tens × 7 = 42 tens
  Regroup as 4 hundreds and 2 tens or 20.
- 6 tens × 80 = 480 tens, or 48 hundreds
  Add the regrouped hundreds: 48 + 4 = 52.

Step 3: Add the partial products.

```
      4
      2
     87
 ×   64
    348
+ 5,220
  [    ]
```

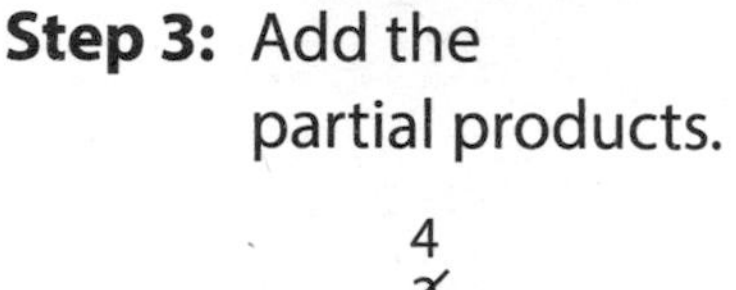

So, 87 × 64 = ________.

Check: Because ________ is close to the estimate, ________, the answer is reasonable.

Show and Grow *I can do it!*

Find the product. Check whether your answer is reasonable.

1. Estimate: ________

```
  41
× 32
```

2. Estimate: ________

```
  □
  5 2
× 4 6
```

3. Estimate: ________

```
  □
  □
  7 8
× 3 5
```

Name ______________________________

Apply and Grow: Practice

Find the product. Check whether your answer is reasonable.

4. Estimate: ________

$$\begin{array}{r} 12 \\ \times\ 43 \\ \hline \end{array}$$

5. Estimate: ________

$$\begin{array}{r} 83 \\ \times\ 24 \\ \hline \end{array}$$

6. Estimate: ________

$$\begin{array}{r} 59 \\ \times\ 76 \\ \hline \end{array}$$

7. Estimate: ________

$22 \times 41 =$ ________

8. Estimate: ________

$94 \times 32 =$ ________

9. Estimate: ________

$63 \times 54 =$ ________

10. Newton eats 14 treats each week. Each treat has 33 calories. How many treat calories does Newton eat each week?

11. **YOU BE THE TEACHER** Your friend finds 43×26. Is he correct? Explain.

$$\begin{array}{r} {}^{1} \\ 43 \\ \times\quad 26 \\ \hline 248 \\ +\quad 860 \\ \hline 1{,}108 \end{array}$$

12. **Open-Ended** Use 4 cards to write 2 two-digit numbers that have a product that is close to, but not greater than, 1,200.

1	2	3
4	5	6
7	8	9

_____ × _____ = _____

Think and Grow: Modeling Real Life

Example There are 16 hours in 1 day on Neptune. There are 88 times as many hours in 1 day on Mercury as 1 day on Neptune. There are 5,832 hours in 1 day on Venus. Are there more hours in 1 day on Mercury or 1 day on Venus?

Multiply to find how many hours there are in 1 day on Mercury.

$$\begin{array}{r} {}^{4} \\ 88 \\ \times \quad 16 \\ \hline 528 \\ + \quad 880 \\ \hline \end{array}$$

☐ hours

Compare.

So, there are more hours in 1 day on ________.

Show and Grow *I can think deeper!*

13. A ninja lanternshark is 18 inches long. A whale shark is 16 times as long as the ninja lanternshark. A hammerhead shark is 228 inches long. Is the whale shark or the hammerhead shark longer?

14. There are 24 science classrooms in a school district. Each classroom receives 3 hot plates. Each hot plate costs $56. How much do all of the hot plates cost?

15. Fourteen adults and 68 students visit the art museum. What is the total cost of admission?

Art Museum Admission Prices	
Adult	$26
Student	$19

Name ______________________________

Homework & Practice 4.6

Multiply two-digit numbers.

Example Find 58×34.

Estimate: $60 \times 35 =$ 2,100

Think: 58 is 5 tens and 8 ones. 34 is 3 tens and 4 ones.

Step 1: Multiply 58 by 4 ones, or 4.

$$\begin{array}{r} {}^{3} \\ 58 \\ \times\ 34 \\ \hline 232 \end{array}$$

$4 \times 58 \rightarrow 232$

Step 2: Multiply 58 by 3 tens, or 30.

$$\begin{array}{r} {}^{2} \\ \not{3} \\ 58 \\ \times\ 34 \\ \hline 232 \\ 1{,}740 \end{array}$$

$30 \times 58 \rightarrow 1{,}740$

Step 3: Add the partial products.

$$\begin{array}{r} {}^{2} \\ \not{3} \\ 58 \\ \times\ 34 \\ \hline 232 \\ +\ 1{,}740 \\ \hline \boxed{1{,}972} \end{array}$$

So, $58 \times 34 =$ 1,972.

Check: Because 1,972 is close to the estimate, 2,100, the answer is reasonable.

Find the product. Check whether your answer is reasonable.

1. Estimate: ________

$$\begin{array}{r} 31 \\ \times\ 92 \\ \hline \end{array}$$

2. Estimate: ________

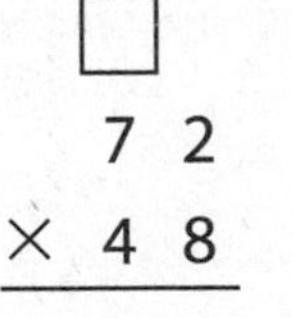

$$\begin{array}{r} 7\ 2 \\ \times\ 4\ 8 \\ \hline \end{array}$$

3. Estimate: ________

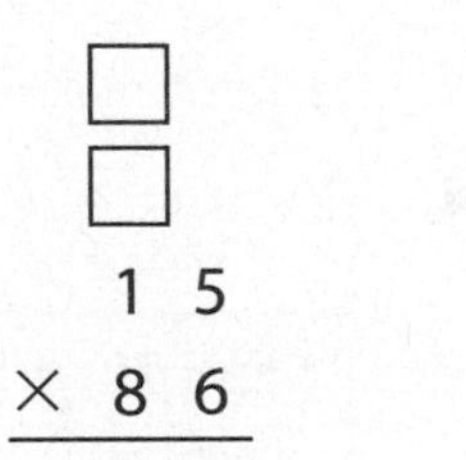

$$\begin{array}{r} 1\ 5 \\ \times\ 8\ 6 \\ \hline \end{array}$$

4. Estimate: ________

$$\begin{array}{r} 81 \\ \times\ 54 \\ \hline \end{array}$$

5. Estimate: ________

$$\begin{array}{r} 23 \\ \times\ 62 \\ \hline \end{array}$$

6. Estimate: ________

$$\begin{array}{r} 97 \\ \times\ 55 \\ \hline \end{array}$$

Find the product. Check whether your answer is reasonable.

7. Estimate: ________

$51 \times 62 =$ ________

8. Estimate: ________

$37 \times 13 =$ ________

9. Estimate: ________

$49 \times 78 =$ ________

10. Newton plays 21 basketball games. He scores 12 points each game. How many points does he score in all?

11. **DIG DEEPER!** When you use regrouping to multiply two-digit numbers, why does the second partial product always end in 0?

12. **MP Number Sense** Find the missing digits.

$$\begin{array}{r} 3\ 4 \\ \times \quad \square\ 5 \\ \hline 1\ \square\ \square \\ +\ 2,0\ 4\ 0 \\ \hline 2,2\ \square\ 0 \end{array}$$

13. **Modeling Real Life** A tiger dives 12 feet underwater. An otter dives 25 times deeper than the tiger. A walrus dives 262 feet underwater. Does the otter or walrus dive deeper?

Review & Refresh

14. Complete the table.

Standard Form	Word Form	Expanded Form
6,835		
		70,000 + 4,000 + 100 + 2
	five hundred one thousand, three hundred twenty-nine	

Name ________________________________

Practice Multiplication Strategies 4.7

Learning Target: Use strategies to multiply two-digit numbers.

Success Criteria:

- I can choose a strategy to multiply.
- I can multiply two-digit numbers.
- I can explain the strategy I used to multiply.

Choose any strategy to find 60×80.

Multiplication Strategies
Place Value
Associative Property of Multiplication
Area Model
Distributive Property
Partial Products
Regrouping

Choose any strategy to find 72×13.

Reasoning Explain why you chose your strategies. Compare your strategies to your partner's strategies. How are they the same or different?

Think and Grow: Practice Multiplication Strategies

Example Find 62×40.

One Way: Use place value.

$62 \times 40 = 62 \times$ ____ tens

$=$ ____ tens

$=$ ____

So, $62 \times 40 =$ ____.

Another Way: Use an area model and partial products.

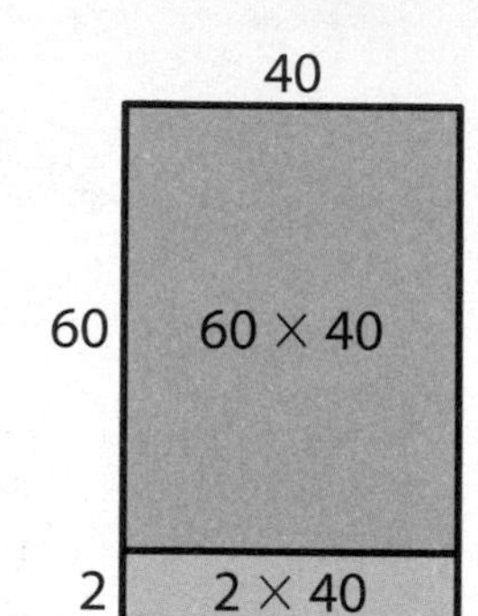

Add the partial products.

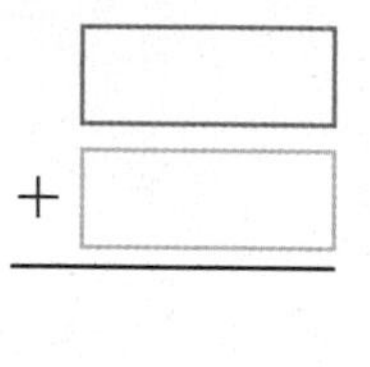

So, $62 \times 40 =$ ____.

Example Find 56×83.

One Way: Use place value and partial products.

$$\begin{array}{r} 56 \\ \times\ 83 \\ \hline \square \quad 80 \times 50 \\ \square \quad 80 \times 6 \\ \square \quad 3 \times 50 \\ + \square \quad 3 \times 6 \\ \hline \square \end{array}$$

So, $56 \times 83 =$ ____.

Another Way: Use regrouping.

Multiply 56 by 3 ones. Then multiply 56 by 8 tens. Regroup if necessary.

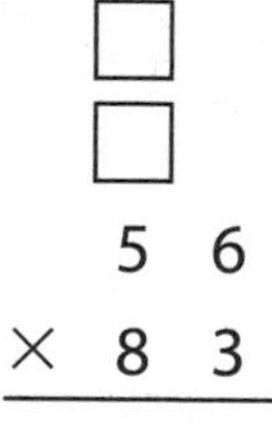

$$\begin{array}{r} 5\ 6 \\ \times\ 8\ 3 \\ \hline \end{array}$$

So, $56 \times 83 =$ ____.

Show and Grow *I can do it!*

Find the product.

1. $90 \times 37 =$ ____

2. $78 \times 21 =$ ____

3. $14 \times 49 =$ ____

Name ______

Apply and Grow: Practice

Find the product.

4. $74 \times 30 =$ ______

5. $51 \times 86 =$ ______

6. $40 \times 29 =$ ______

7. $92 \times 80 =$ ______

8. $41 \times 17 =$ ______

9. $60 \times 53 =$ ______

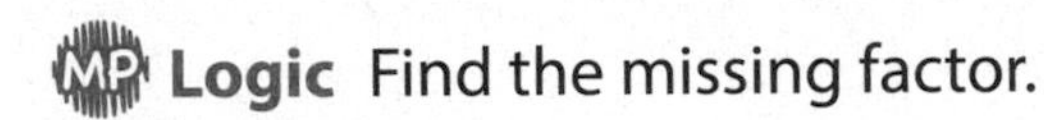

Logic Find the missing factor.

10.
$$\begin{array}{r} 72 \\ \times \square \\ \hline 72 \\ +\ 2{,}880 \\ \hline 2{,}952 \end{array}$$

11.
$$\begin{array}{r} 65 \\ \times \square \\ \hline 260 \\ +\ 1{,}950 \\ \hline 2{,}210 \end{array}$$

12.
$$\begin{array}{r} 93 \\ \times \square \\ \hline 651 \\ +\ 7{,}440 \\ \hline 8{,}091 \end{array}$$

13. **Writing** Explain why you start multiplying with the ones place when using regrouping to multiply.

14. **DIG DEEPER!** Find the missing digit so that both products are the same.

$$\begin{array}{r} 26 \\ \times 15 \\ \hline \end{array} \qquad \begin{array}{r} 3\ 0 \\ \times\ 1\ \square \\ \hline \end{array}$$

Think and Grow: Modeling Real Life

Example A swinging ship ride runs 50 times each afternoon. The ship has 10 rows of benches with 4 seats in each bench. If the ship is full each time it runs, how many people will ride the swinging ship in 1 afternoon?

Multiply to find how many people will ride the swinging ship ride each time.

$10 \times 4 =$ ______

______ people will ride the swinging ship ride each time.

Multiply to find how many people will ride the swinging ship in 1 afternoon.

$40 \times 50 = 40 \times (5 \times 10)$ Rewrite 50 as 5×10.

$= (40 \times 5) \times 10$ Associative Property of Multiplication

$=$ ______ $\times 10$

$=$ ______

______ people will ride the swinging ship in 1 afternoon.

Show and Grow *I can think deeper!*

15. A teacher orders 25 rock classification kits. Each kit has 4 rows with 9 rocks in each row. How many rocks are there in all?

16. A hotel has 12 floors with 34 rooms on each floor. 239 rooms are in use. How many rooms are *not* in use?

17. A child ticket for a natural history museum costs \$13. An adult ticket costs twice as much as a child ticket. How much does it cost for 21 children and 14 adults to go to the museum?

Name ______________________

Homework & Practice 4.7

Learning Target: Use strategies to multiply two-digit numbers.

Example Find 17×80.

Use the Associative Property of Multiplication.

$17 \times 80 = 17 \times (8 \times 10)$

$= (17 \times 8) \times 10$

$= 136 \times 10$

$= 1{,}360$

So, $17 \times 80 =$ <u>1,360</u>.

Example Find 34×52.

Use the Distributive Property.

$34 \times 52 = (30 + 4) \times (50 + 2)$

$= (30 + 4) \times 50 + (30 + 4) \times 2$

$= (30 \times 50) + (4 \times 50) + (30 \times 2) + (4 \times 2)$

$= 1{,}500 + 200 + 60 + 8$

$= 1{,}768$

So, $34 \times 52 =$ <u>1,768</u>.

Find the product.

1. $16 \times 13 =$ ______

2. $29 \times 50 =$ ______

3. $78 \times 45 =$ ______

4. $30 \times 71 =$ ______

5. $62 \times 14 =$ ______

6. $80 \times 90 =$ ______

Find the product.

7. $70 \times 18 =$ ______

8. $32 \times 59 =$ ______

9. $67 \times 20 =$ ______

10. $51 \times 84 =$ ______

11. $40 \times 40 =$ ______

12. $23 \times 97 =$ ______

13. **Writing** Which strategy do you prefer to use when multiplying two-digit numbers? Explain.

14. **MP Patterns** What number can you multiply the number of tires by to find the total weight? Use this pattern to complete the table.

Number of Tires	4	8	12	16	20
Total Weight (pounds)	80	160	240		

15. **Modeling Real Life** Each bag of popcorn makes 13 cups. A school has a movie day, and the principal brings 15 boxes of popcorn. Each box has 3 bags of popcorn. How many cups of popcorn does the principal bring?

Review & Refresh

Find the product.

16. $8 \times 200 =$ ______

17. $7 \times 300 =$ ______

18. $6{,}000 \times 5 =$ ______

19. $9 \times 90 =$ ______

20. $3{,}000 \times 6 =$ ______

21. $5 \times 500 =$ ______

Name ________________________________

Problem Solving: Multiplication with Two-Digit Numbers

4.8

Learning Target: Solve multi-step word problems involving two-digit multiplication.

Success Criteria:

- I can understand a problem.
- I can make a plan to solve using letters to represent the unknown numbers.
- I can solve a problem using an equation.

Explore and Grow

Explain, in your own words, what the problem below is asking. Then explain how you can use multiplication to solve the problem.

A ferry can transport 64 cars each time it leaves a port. The ferry leaves a port 22 times in 1 day. How many cars can the ferry transport in 1 day?

Construct Arguments Make a plan to find how many cars the ferry can transport in 1 week.

Think and Grow: Problem Solving: Multiplication with Two-Digit Numbers

Example A pet store receives a shipment of 8 boxes of dog treats. Each box is 2 feet high and has 18 bags of dog treats. How many ounces of dog treats does the pet store receive in the shipment?

Understand the Problem

What do you know?

- The store receives 8 boxes.
- Each box is 2 feet high.
- Each box has 18 bags of dog treats.
- Each bag weighs 32 ounces.

What do you need to find?

- You need to find how many ounces of dog treats the pet store receives in the shipment.

Make a Plan

How will you solve?

- Multiply 32 by 18 to find how many ounces of dog treats are in each box.
- Then multiply the product by 8 to find how many ounces of dog treats the pet store receives in the shipment.
- The height of each box is unnecessary information.

Solve

Step 1: How many ounces of dog treats are in each box?

$32 \times 18 = b$

b is the unknown product.

$$\begin{array}{r} \square \\ 3\ 2 \\ \times\ 1\ 8 \\ \hline \end{array}$$

$b =$ ______

Step 2: Use b to find how many ounces of dog treats the pet store receives.

$b \times 8 = p$

p is the unknown product.

$$\begin{array}{r} \square \\ \times\quad 8 \\ \hline \end{array}$$

$p =$ ______

The pet store receives ______ ounces of dog treats.

Show and Grow *I can do it!*

1. Show how to solve the problem above using one equation.

Name ______________________________

Apply and Grow: Practice

Understand the problem. What do you know? What do you need to find? Explain.

2. Thirteen students create a petition for longer recess. They need 5,000 signatures in all. So far, each student has 99 signatures. How many more signatures do they need?

3. An activity book has 35 pages and costs $7. Each page has 4 puzzles. You have completed all of the puzzles on 16 of the pages. How many puzzles do you have left to complete?

Understand the problem. Then make a plan. How will you solve? Explain.

4. Twelve classes provide items for a time capsule. There are 23 students in each class. Each student puts 2 small items in the time capsule. How many items are in the time capsule?

5. A craftsman cuts letters and numbers from license plates to make signs. He has 15 Florida plates and 25 Georgia plates. Each plate has a total of 7 letters and numbers. How many letters and numbers does the craftsman cut in all?

6. **DIG DEEPER!** In 1 month, the solar panel and the wind turbine can produce the kilowatt hours of electricity shown. How much electricity can 28 solar panels and 1 small wind turbine produce each month?

Wind turbine:
833 kilowatt hours

Solar panel:
30 kilowatt hours

Think and Grow: Modeling Real Life

Example An adult ticket for a zipline course costs \$48. A child ticket costs \$19 less than an adult ticket. In 1 day, 86 adults and 42 children ride the zipline. How much more money was earned from adult tickets than from child tickets?

Think: What do you know? What do you need to find? How will you solve?

Step 1: How much money was earned from adult tickets?

$\$48 \times 86 = a$

a is the unknown product.

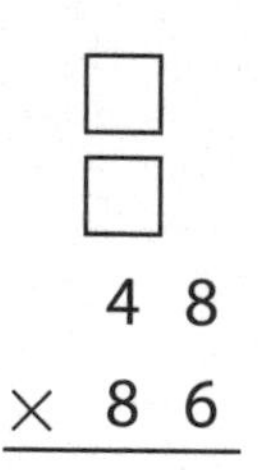

$$\begin{array}{r} 48 \\ \times\ 86 \\ \hline \end{array}$$

$a =$ ______

Step 2: How much money was earned from child tickets?

A child ticket costs $\$48 - \$19 = \$29$.

$\$29 \times 42 = c$

c is the unknown product.

$$\begin{array}{r} \square \\ \square \\ 29 \\ \times\ 42 \\ \hline \end{array}$$

$c =$ ______

Step 3: Use m to represent how much more money was earned from adult tickets.

$c =$ ______	m

$a =$ ______

$$\begin{array}{r} \square \\ -\ \square \\ \hline \end{array}$$

m is the unknown difference.

$m =$ ______

So, \$__________ more was earned from adult tickets.

Show and Grow *I can think deeper!*

7. A theater has 45 rows of 72 seats on the floor level and 22 rows of 36 seats in the balcony. How many seats are there in all?

How many more seats are on the floor level than in the balcony?

Name ______________________

Homework & Practice 4.8

Learning Target: Solve multi-step word problems involving two-digit multiplication.

Example A store receives a shipment of 5 boxes of pretzels. Each box is 50 centimeters high and has 24 bags of pretzels. How many ounces of pretzels does the store receive in the shipment?

Think: What do you know? What do you need to find? How will you solve?

Step 1: How many ounces of pretzels are in each box?

$16 \times 24 = b$

b is the unknown product.

```
   1
   2
   1 6
×  2 4
   6 4
+ 3 2 0
  3 8 4
```

$b =$ 384

Step 2: Use b to find how many ounces of pretzels the store receives.

384 $\times 5 = p$

p is the unknown product.

```
   42
  384
×   5
1,920
```

$p =$ 1,920

The store receives 1,920 ounces of pretzels.

Understand the problem. Then make a plan. How will you solve? Explain.

1. Seventy-two mushers compete in a sled-dog race. Each musher has 16 dogs. How many more dogs compete in the race than mushers?

2. A photographer buys 3 USB drives that each cost \$5. She puts 16 folders on each drive. Each folder has 75 photographs. How many photographs does the photographer put on the USB drives in all?

3. A teacher has 68 students take a 25-question test. The teacher checks the answers for 9 of the tests. How many answers does the teacher have left to check?

4. Each day, a cyclist bikes uphill for 17 miles and downhill for 18 miles. She drinks 32 fluid ounces of water after each bike ride. How many miles does the cyclist bike in 2 weeks?

5. MP **Precision** Which expressions can be used to solve the problem?

Twelve friends play a game that has 308 cards. Each player receives 16 cards. How many cards are left?

$(308 - 12) \times 16$ $\quad$ $308 - (16 \times 12)$

$308 - (12 \times 16)$ $\quad$ $(308 - 16) - 12$

6. **Modeling Real Life** A child ticket costs $14 less than an adult ticket. What is the total ticket cost for 18 adults and 37 children?

7. **Modeling Real Life** An artist creates a pattern by alternating square and rectangular tiles. The pattern has 14 square tiles and 13 rectangular tiles. How long is the pattern?

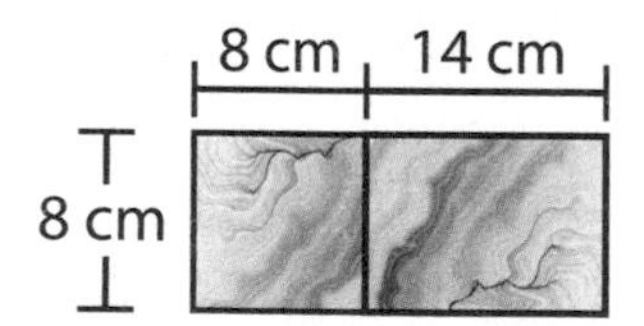

8. **Modeling Real Life** A cargo ship has 34 rows of crates. Each row has 16 stacks of crates. There are 5 crates in each stack. The ship workers unload 862 crates. How many crates are still on the ship?

Review & Refresh

Estimate the product.

9. 4×85

10. 6×705

11. $8 \times 7{,}923$

Name ____________________

Performance Task

Wind turbines convert wind to energy. Most wind turbines have 3 blades. The blades rotate slower or faster depending on the speed of the wind. More energy is generated when the blades spin faster.

1. A wind turbine rotates between 15 and 40 times in 1 minute.

a. What is the least number of times the turbine rotates in 1 hour?

b. What is the greatest number of times the turbine rotates in 1 hour?

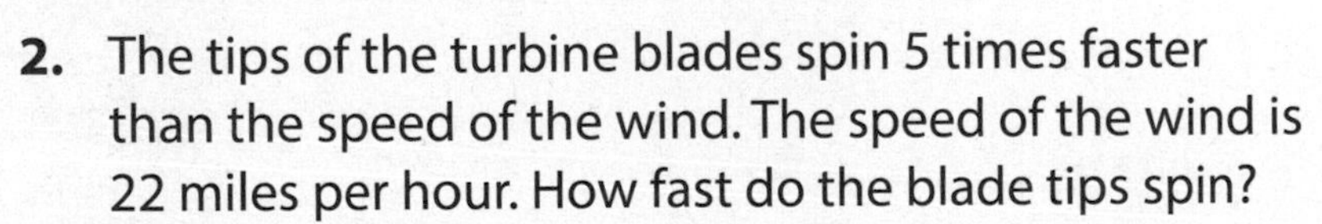

2. The tips of the turbine blades spin 5 times faster than the speed of the wind. The speed of the wind is 22 miles per hour. How fast do the blade tips spin?

3. A turbine farm has 7 large wind turbines. Each wind turbine can generate enough energy to power 1,485 houses. How many houses can the turbine farm power in all?

Wind Turbines

Length of Each Blade (meters)	Number of Houses Powered
15	110
30	440

4. Use the chart.

a. How many times more houses are powered when the length of each blade is doubled?

b. A wind turbine has blades that are each 60 meters long. How many houses can the wind turbine power?

Multiplication Boss

Directions:

1. Each player flips 4 Number Cards and uses them in any order to create a multiplication problem with two-digit factors.
2. Each player finds the product of the two factors.
3. Players compare products. The player with the greater product takes all 8 cards.
4. If the products are equal, each player flips 4 more cards and plays again. The player with the greater product takes all 16 cards.
5. The player with the most cards at the end of the round wins!

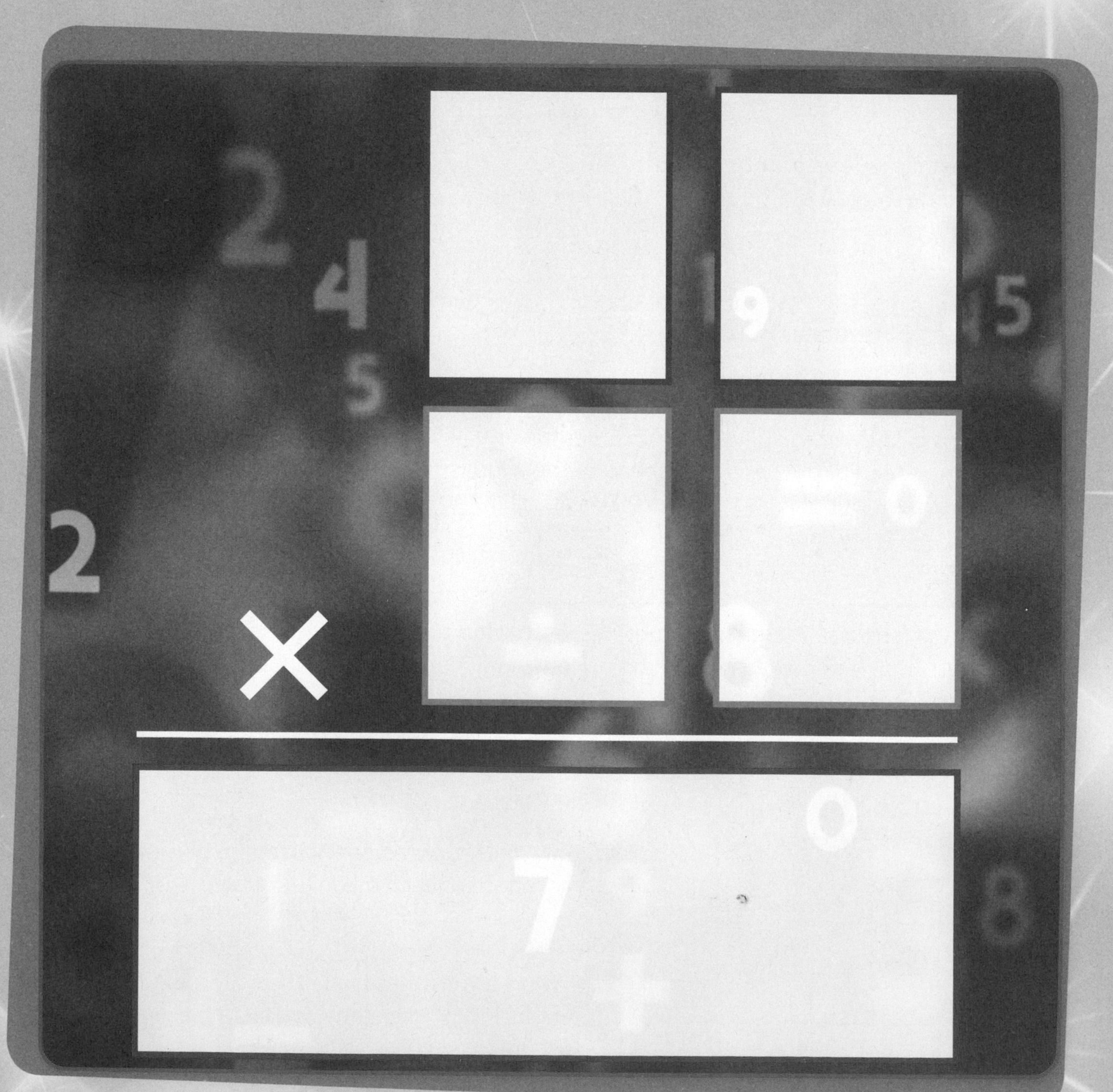

Name ______________________________

Chapter Practice

4.1 Multiply by Tens

Find the product.

1. $50 \times 20 =$ ______

2. $30 \times 60 =$ ______

3. $80 \times 10 =$ ______

4. $40 \times 70 =$ ______

5. $60 \times 50 =$ ______

6. $90 \times 90 =$ ______

7. $70 \times 11 =$ ______

8. $18 \times 30 =$ ______

9. $20 \times 75 =$ ______

Find the missing factor.

10. $40 \times$ ______ $= 3{,}200$

11. ______ $\times 20 = 1{,}200$

12. $30 \times$ ______ $= 2{,}100$

4.2 Estimate Products

Estimate the product.

13. 25×74

14. 16×28

15. 42×81

Open-Ended Write two possible factors that can be estimated as shown.

16. 8,100

______ × ______

↓ ↓

______ × ______ = 8,100

17. 400

______ × ______

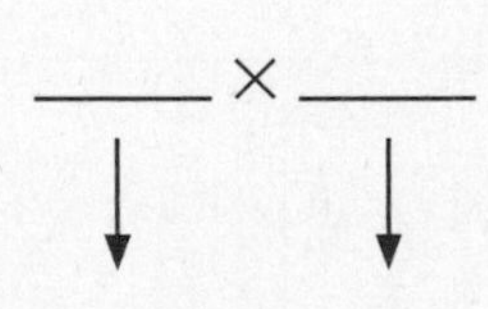

______ × ______ = 400

4.3 Use Area Models to Multiply Two-Digit Numbers

Draw an area model to find the product.

18. $13 \times 19 =$ ________

19. $21 \times 36 =$ ________

20. **YOU BE THE TEACHER** Your friend finds 28×24. Is your friend correct? Explain.

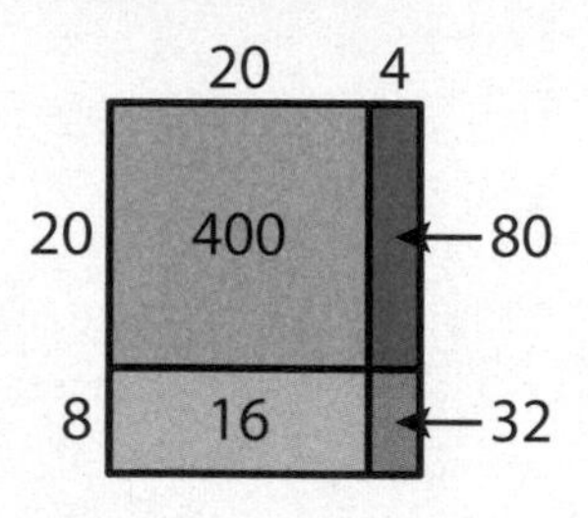

$400 + 16 + 80 + 32 = 528$

4.4 Use Distributive Property to Multiply Two-Digit Numbers

Use the Distributive Property to find the product.

21. $27 \times 34 = 27 \times (30 + 4)$

$= (27 \times 30) + (27 \times 4)$

$= (20 + 7) \times$ ______ $+ (20 + 7) \times$ ______

$= (20 \times$ ______$) + (7 \times$ ______$) + (20 \times$ ______$) + (7 \times$ ______$)$

$=$ ______ $+$ ______ $+$ ______ $+$ ______

$=$ ______

So, $27 \times 34 =$ ______.

22. $43 \times 18 =$ ________

23. $35 \times 57 =$ ________

24. $81 \times 76 =$ ________

4.5 Use Partial Products to Multiply Two-Digit Numbers

Find the product. Check whether your answer is reasonable.

25. $\begin{array}{r} 18 \\ \times\ 22 \\ \hline \end{array}$

26. $\begin{array}{r} 73 \\ \times\ 46 \\ \hline \end{array}$

27. $\begin{array}{r} 39 \\ \times\ 84 \\ \hline \end{array}$

28. $57 \times 19 =$ ______

29. $38 \times 65 =$ ______

30. $94 \times 26 =$ ______

MP Reasoning Find the missing digits. Then find the product.

31.

$$\begin{array}{r} \square 2 \\ \times\ \square 5 \\ \hline 300 \\ 60 \\ 50 \\ +\ 10 \\ \hline \square \end{array}$$

32.

$$\begin{array}{r} \square 1 \\ \times\ \square 4 \\ \hline 2{,}000 \\ 40 \\ 200 \\ +\ 4 \\ \hline \square \end{array}$$

4.6 Multiply Two-Digit Numbers

Find the product. Check whether your answer is reasonable.

33. Estimate: ______

$\begin{array}{r} 62 \\ \times\ 34 \\ \hline \end{array}$

34. Estimate: ______

$\begin{array}{r} 87 \\ \times\ 91 \\ \hline \end{array}$

35. Estimate: ______

$\begin{array}{r} 73 \\ \times\ 45 \\ \hline \end{array}$

36. Estimate: ______

$13 \times 21 =$ ______

37. Estimate: ______

$42 \times 53 =$ ______

38. Estimate: ______

$29 \times 66 =$ ______

4.7 Practice Multiplication Strategies

Find the product.

39. $80 \times 30 =$ ______

40. $26 \times 51 =$ ______

41. $94 \times 70 =$ ______

42. $15 \times 67 =$ ______

43. $40 \times 38 =$ ______

44. $29 \times 92 =$ ______

45. Modeling Real Life A Ferris wheel runs 40 times each day. It has 16 cars with 4 seats in each car. If the Ferris wheel is full each time it runs, how many people will ride it in 1 day?

4.8 Problem Solving: Multiplication with Two-Digit Numbers

46. A music fan memorizes 59 songs for a concert. Her goal is to memorize all of the songs from 13 albums. There are 15 songs on each album. How many more songs does the music fan still need to memorize?

47. Find the area of the Jamaican flag.

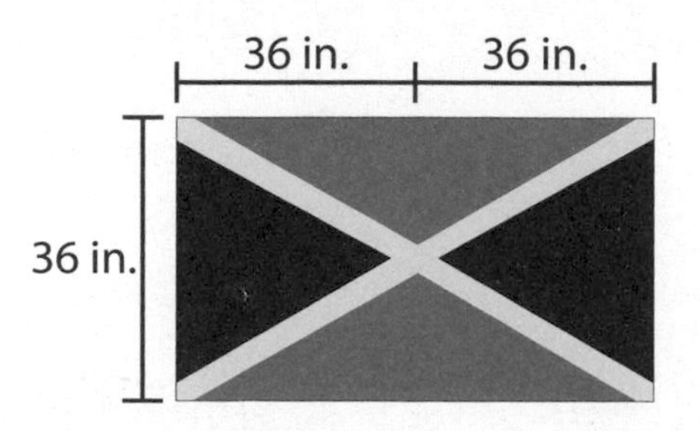

Divide Multi-Digit Numbers by One-Digit Numbers

Chapter Learning Target:
Understand dividing one-digit numbers.

Chapter Success Criteria:
- I can divide a number.
- I can use division facts to estimate a quotient.
- I can write division problems.
- I can solve division problems.

- What is a planetarium?
- Your grade is going on a field trip to a planetarium. How can you use division to estimate the number of people in each tour group?

Name ______________________________

5 Vocabulary

Review Words
dividend
divisor
quotient

Organize It

Use the review words to complete the graphic organizer.

[ ] 10 ÷ 2 = 5 The answer when you divide one number by another number	[ ] 10 ÷ 2 = 5 The number of objects or the amount you want to divide
[ ] 10 ÷ 2 = 5 The number by which you divide	Model 10 ÷ 2 = 5 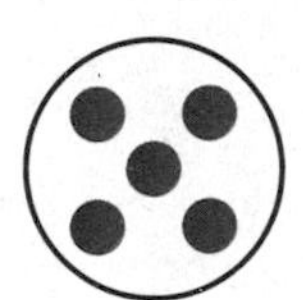 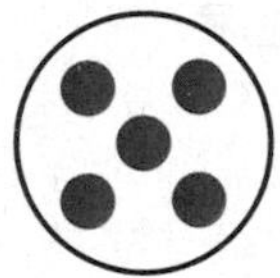

Division

Define It

Use your vocabulary cards to match.

1. partial quotients

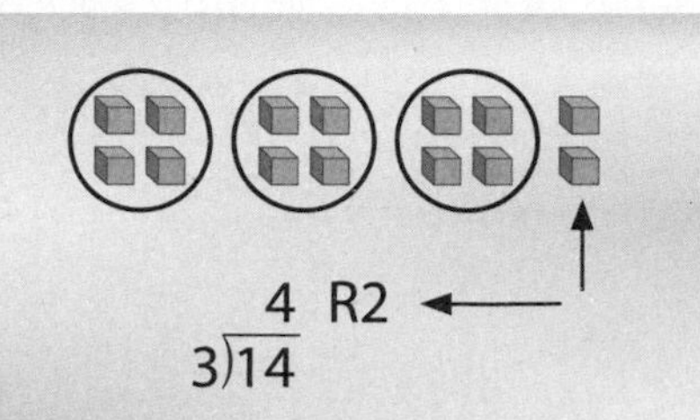

2. remainder

$$\begin{array}{r} 6\overline{)84} \\ -\ 60 = 6 \times 10 \\ \hline 24 \\ -\ 24 = 6 \times 4 \\ \hline 0 \end{array} \qquad \begin{array}{r} 10 \\ +\ 4 \\ \hline 14 \end{array}$$

Chapter 5 Vocabulary Cards

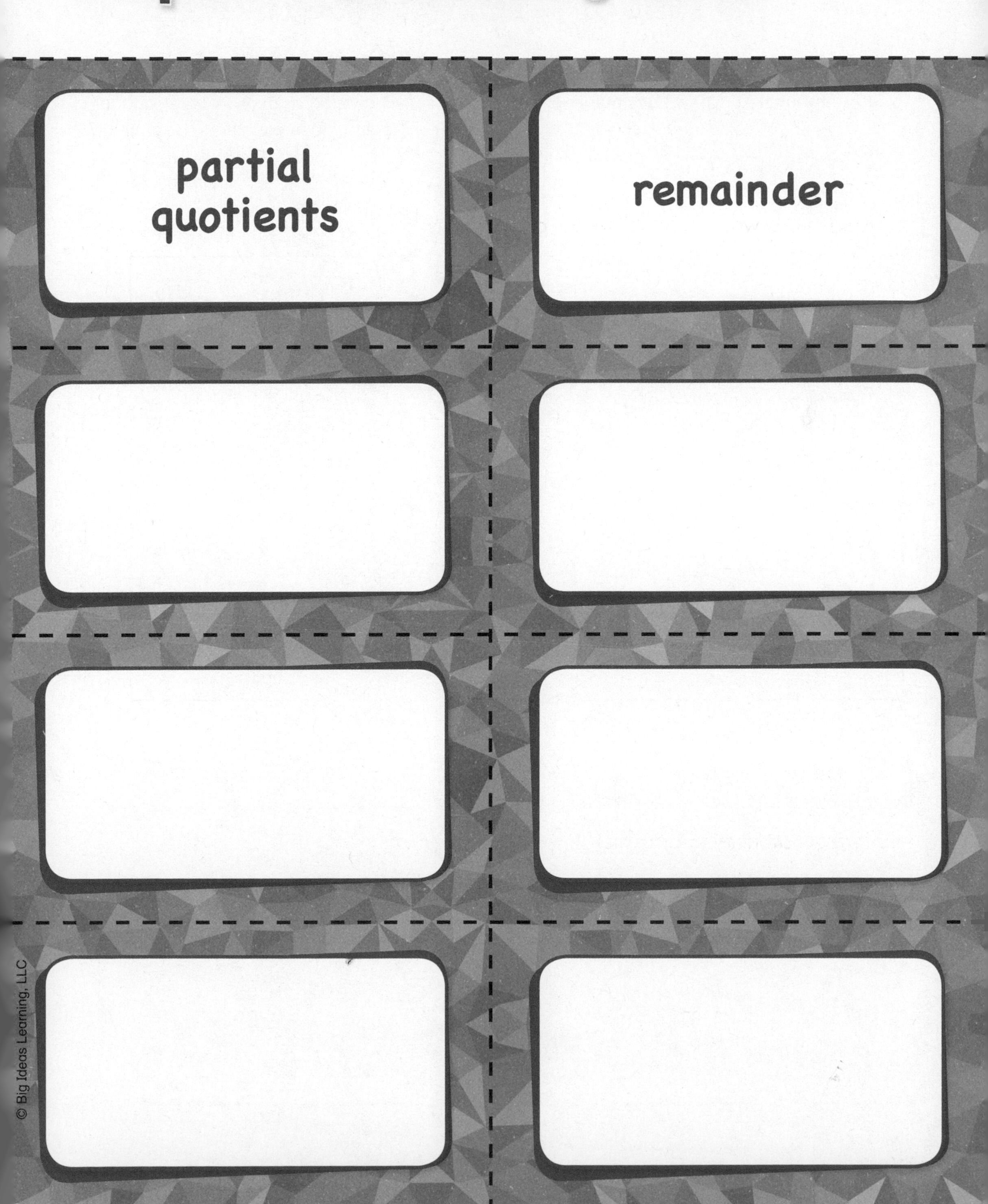

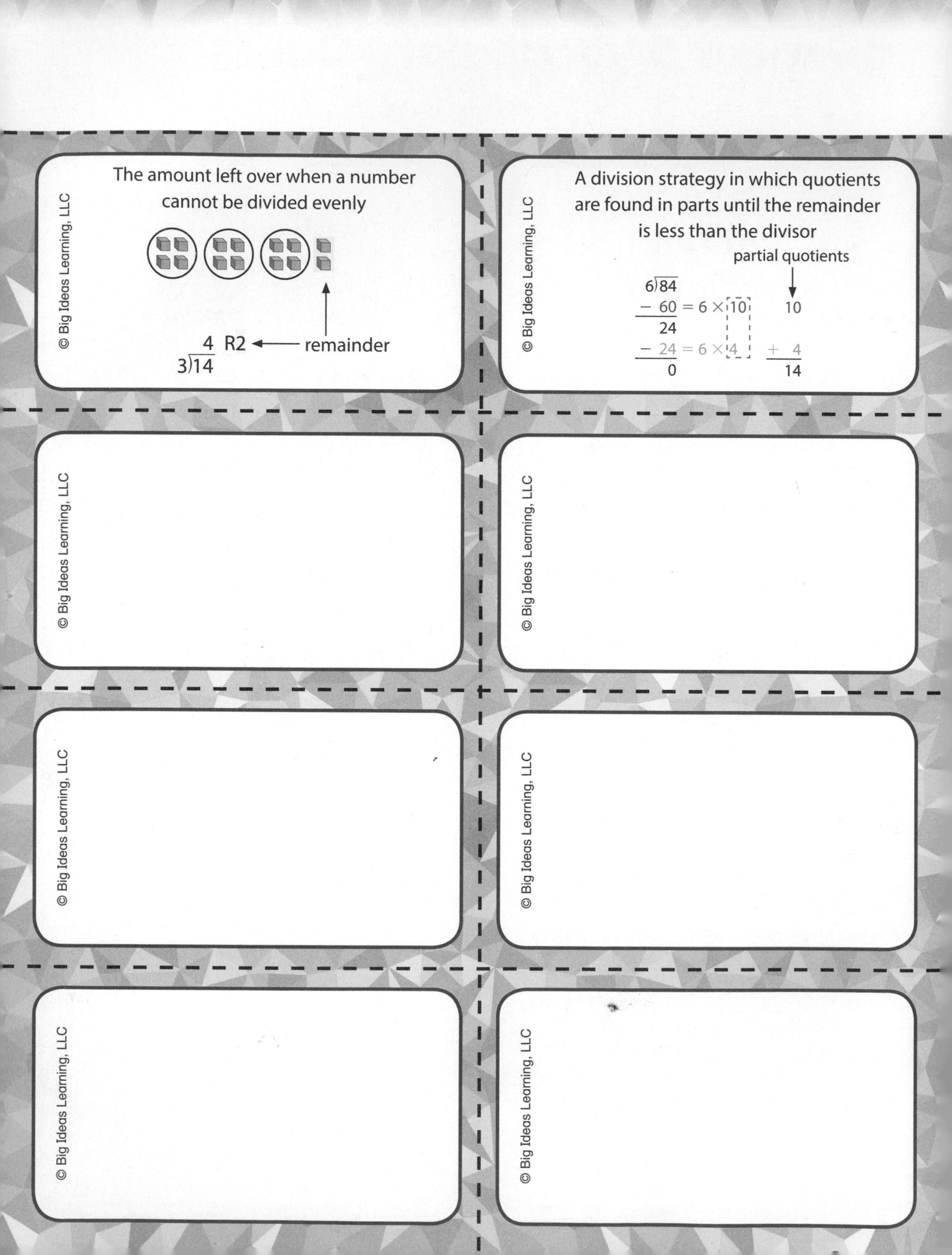

The amount left over when a number cannot be divided evenly
4 R2
3)14
remainder
© Big Ideas Learning, LLC
A division strategy in which quotients are found in parts until the remainder is less than the divisor
partial quotients
6)84
− 60 = 6 × 10
10
24
− 24 = 6 × 4
+ 4
0
14
© Big Ideas Learning, LLC
© Big Ideas Learning, LLC
© Big Ideas Learning, LLC
© Big Ideas Learning, LLC
© Big Ideas Learning, LLC
© Big Ideas Learning, LLC
© Big Ideas Learning, LLC

Name ______________________

Divide Tens, Hundreds, and Thousands 5.1

Learning Target: Use place value to divide tens, hundreds, or thousands.

Success Criteria:

- I can divide a multiple of ten, one hundred, or one thousand by a one-digit number.
- I can explain how to use place value and division facts to divide tens, hundreds, or thousands.

Explore and Grow

Use a model to find each missing factor. Draw each model. Then write the related division equation.

14 $\times 2 = 8$	40 $\times 2 = 80$
400 $\times 2 = 800$	4,000 $\times 2 = 8{,}000$

What pattern do you notice?

Repeated Reasoning Explain how $12 \div 4$ can help you find $1{,}200 \div 4$.

12 ÷ 4 = 3

4 x 3 ③

1,200 ÷ 4 = 300

300

Think and Grow: Divide Tens, Hundreds, and Thousands

You can use place value and basic division facts to divide tens, hundreds, or thousands by one-digit numbers.

Example Find 270 ÷ 9.

3	3	3	3	3	3	3	3	3

27

Think: 27 ÷ 9 = 3 Division fact

30	30	30	30	30	30	30	30	30

270

270 ÷ 9 = 30 tens ÷ 9 Use place value.

= 3 tens Divide.

= 30

So, 270 ÷ 9 = ______.

Example Find 5,600 ÷ 8.

Think: 56 ÷ 8 = 7 Division fact

5,600 ÷ 8 = 56 hundreds ÷ 8 Use place value.

= 7 hundreds Divide.

= 700

So, 5,600 ÷ 8 = ______.

Show and Grow *I can do it!*

1. Find 2,400 ÷ 6.

Think: 24 ÷ 6 = 4

2,400 ÷ 6 = 24 hundreds ÷ 6

= 4 hundreds

= 400

So, 2,400 ÷ 6 = 400.

2. Find each quotient.

49 ÷ 7 = 7

490 ÷ 7 = 70

4,900 ÷ 7 = 700

Name ___________________________

Apply and Grow: Practice

Find the quotient.

3. $50 \div 5 =$ ____

4. $360 \div 6 =$ ____

5. $7{,}200 \div 8 =$ ____

6. $180 \div 2 =$ ____

7. $4{,}200 \div 7 =$ ____

8. $20 \div 2 =$ ____

9. $2{,}000 \div 5 =$ ____

10. $30 \div 3 =$ ____

11. $320 \div 4 =$ ____

12. $140 \div 2 =$ ____

13. $5{,}400 \div 9 =$ ____

14. $180 \div 6 =$ ____

DIG DEEPER! Find the missing number.

15. $70 \div$ ____ $= 10$

16. $4{,}000 \div$ ____ $= 800$

17. $160 \div$ ____ $= 40$

18. ____ $\div 7 = 300$

19. ____ $\div 5 = 70$

20. ____ $\div 6 = 10$

Compare.

21. $40 \div 4$ ◯ 1×10

22. $160 \div 8$ ◯ 2×100

23. $8{,}100 \div 9$ ◯ 9×10

24. There are 240 students visiting a fair. They are divided equally among 8 barns. How many students are in each barn?

25. **YOU BE THE TEACHER** Is Descartes correct? Explain.

Think and Grow: Modeling Real Life

Example A lobster lays 5,400 eggs. It lays 9 times as many eggs as a seahorse. How many eggs does the seahorse lay?

Draw a model.

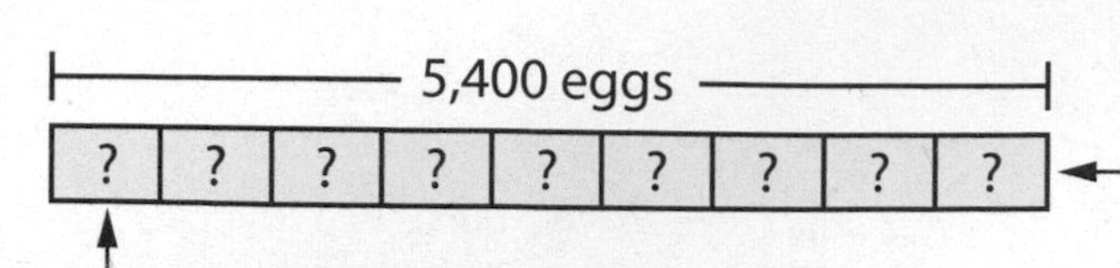

The lobster lays 9 times as many eggs as the seahorse.

The ? represents how many eggs the seahorse lays.

Divide 5,400 by 9 to find how many eggs the seahorse lays.

Think: $54 \div 9 =$ _____ Division fact

$5{,}400 \div 9 =$ _____ hundreds $\div 9$ Use place value.

$=$ _____ hundreds Divide.

$=$ _____ So, the seahorse lays _____ eggs.

Show and Grow *I can think deeper!*

26. A coach has 350 career wins. He has 7 times as many career wins as wins this season. How many wins does the coach have this season?

27. There are 160 shareable bicycles in a city. There are 8 bicycle-sharing stations. Each station has the same number of bicycles. How many bicycles are at each station?

28. A charity has 637 adult volunteers and 563 teenage volunteers. All of the volunteers are divided into 6 equal groups. How many volunteers are in each group?

Name ________________________________

Estimate Quotients 5.2

Learning Target: Use division facts and compatible numbers to estimate quotients.

Success Criteria:

- I can use division facts and compatible numbers to estimate a quotient.
- I can find two estimates that a quotient is between.

Explain how you can use the table to estimate $740 \div 8$.

$10 \times 8 =$ _____	$60 \times 8 =$ _____
$20 \times 8 =$ _____	$70 \times 8 =$ _____
$30 \times 8 =$ _____	$80 \times 8 =$ _____
$40 \times 8 =$ _____	$90 \times 8 =$ _____
$50 \times 8 =$ _____	$100 \times 8 =$ _____

$740 \div 8$ is about _____.

Reasoning Why did you choose your estimate? Compare your results with your partner.

Think and Grow: Estimate Quotients

You can use division facts and compatible numbers to estimate a quotient.

Example Estimate 154 ÷ 4.

Look at the first two digits of the dividend and use basic division facts.

Think: What number close to 154 is easily divided by 4?

Try 120. 12 ÷ 4 = ______, so 120 ÷ 4 = ______.

Try 160. 16 ÷ 4 = ______, so 160 ÷ 4 = ______.

Choose 160 because 154 is closer to 160.

So, 154 ÷ 4 is about ______.

When solving division problems, you can check whether an answer is reasonable by finding two numbers that a quotient is between.

Example Find two numbers that the quotient 6,427 ÷ 7 is between.

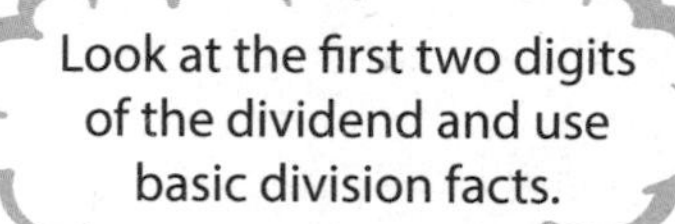

Think: What numbers close to 6,427 are easily divided by 7?

Use 6,300. 63 ÷ 7 = ______, so 6,300 ÷ 7 = ______.

Use 7,000. 70 ÷ 7 = ______, so 7,000 ÷ 7 = ______.

6,427 is between 6,300 and 7,000.

So, the quotient 6,427 ÷ 7 is between ______ and ______.

Show and Grow *I can do it!*

Estimate the quotient.

1. 61 ÷ 3

2. 465 ÷ 9

Find two numbers that the quotient is between.

3. 477 ÷ 8

4. 5,194 ÷ 6

Name ________________________________

Apply and Grow: Practice

Estimate the quotient.

5. $29 \div 5$

6. $571 \div 8$

7. $202 \div 6$

8. $3,384 \div 7$

Find two estimates that the quotient is between.

9. $22 \div 3$

10. $165 \div 9$

11. $2,387 \div 5$

12. $3,813 \div 4$

DIG DEEPER! Estimate to compare.

13. $26 \div 9$ ◯ 2

14. $142 \div 3$ ◯ 50

15. $645 \div 8$ ◯ $816 \div 9$

16. $1,200 \div 6$ ◯ $800 \div 4$

17. A pizza shop owner has 2,532 coupons in pamphlets of 4 coupons each. He wants to determine whether he has enough pamphlets to give one to each of his first 600 customers. Can he use an estimate, or is an exact answer required? Explain.

YOU BE THE TEACHER Your friend finds the quotient. Is his answer reasonable? Estimate to check.

18. $273 \div 3 \stackrel{?}{=} 91$

19. $4,290 \div 6 \stackrel{?}{=} 615$

Think and Grow: Modeling Real Life

The Taal Volcano is the smallest active volcano on Earth.

Example Mount Nantai is 2,486 meters above sea level. It is about 8 times as many meters above sea level as the Taal Volcano. About how many meters above sea level is the Taal Volcano?

Mount Nantai is about 8 times as many meters above sea level as the Taal Volcano, so estimate 2,486 ÷ 8.

Think: What number close to 2,486 is easily divided by 8?

Try 2,400. 24 ÷ 8 = ______, so 2,400 ÷ 8 = ______.

Try 3,200. 32 ÷ 8 = ______, so 3,200 ÷ 8 = ______.

Choose 2,400 because 2,486 is closer to 2,400.

So, the Taal Volcano is about __________ meters above sea level.

Show and Grow *I can think deeper!*

20. There are about 3,785 milliliters in 1 gallon. There are 4 times as many milliliters in 1 gallon as there are in 1 quart. About how many milliliters are in 1 quart?

21. A teenager works at an amusement park for 3 months and earns $2,178. She earns the same amount each month. About how much money does she earn each month?

22. An animal shelter has a bin filled with 456 pounds of dog food. There are 4 large dogs at the shelter who each eat 2 pounds of the dog food each day. For about how many days can the dogs eat from the bin of food?

Name ____________________

Homework & Practice 5.2

Learning Target: Use division facts and compatible numbers to estimate quotients.

Example Estimate 3,587 ÷ 6.

Think: What number close to 3,587 is easily divided by 6?

Try 3,000. 30 ÷ 6 = __5__, so 3,000 ÷ 6 = __500__.

Try 3,600. 36 ÷ 6 = __6__, so 3,600 ÷ 6 = __600__.

Choose 3,600 because 3,587 is closer to 3,600.

So, 3,587 ÷ 6 is about __600__.

Example Find two numbers that the quotient 294 ÷ 3 is between.

Think: What numbers close to 294 are easily divided by 3?

Use 270. 27 ÷ 3 = __9__, so 270 ÷ 3 = __90__.

Use 300. 30 ÷ 3 = __10__, so 300 ÷ 3 = __100__.

294 is between 270 and 300.

So, the quotient of 294 ÷ 3 is between __90__ and __100__.

Estimate the quotient.

1. 33 ÷ 4

2. 527 ÷ 9

Find two estimates that the quotient is between.

3. 308 ÷ 7

4. 3,421 ÷ 6

DIG DEEPER! Estimate to compare.

5. $97 \div 3 \bigcirc 40$

6. $425 \div 5 \bigcirc 182 \div 7$

7. Three friends want to share 261 tickets equally. They want to determine whether they can each have at least 87 tickets. Can they use an estimate, or is an exact answer required? Explain.

8. **MP Reasoning** Explain how to find a better estimate for $462 \div 5$ than the one shown.

Round 462 to 500. Estimate $500 \div 5$.
$500 \div 5 = 100$, so $462 \div 5$ is about 100.

9. **Modeling Real Life** A machine that makes toy spinners is in operation for 8 hours each day. The machine makes 7,829 toy spinners in 1 day. About how many toy spinners does the machine make each hour?

10. **Modeling Real Life** A little penguin has 10,235 feathers. The penguin has about 3 times as many feathers as a blue jay. About how many feathers does the blue jay have?

Review & Refresh

Write an equation for the comparison sentence.

11. 15 is 9 more than 6.

12. 56 is 7 times as many as 8.

Name ______________________

Understand Division and Remainders 5.3

Learning Target: Use models to find quotients and remainders.

Success Criteria:

- I can use models to divide numbers that do not divide evenly.
- I can find a quotient and a remainder.
- I can interpret the quotient and the remainder in a division problem.

Explore and Grow

Use base ten blocks to determine whether 14 can be divided equally among 2, 3, 4, or 5 groups. Draw and describe your models.

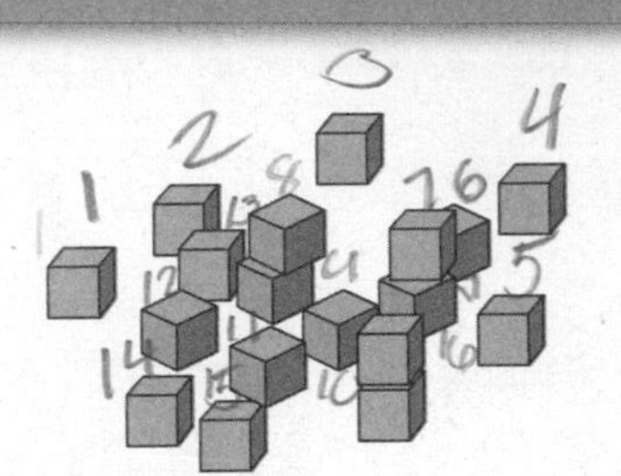

2 equal groups	3 equal groups
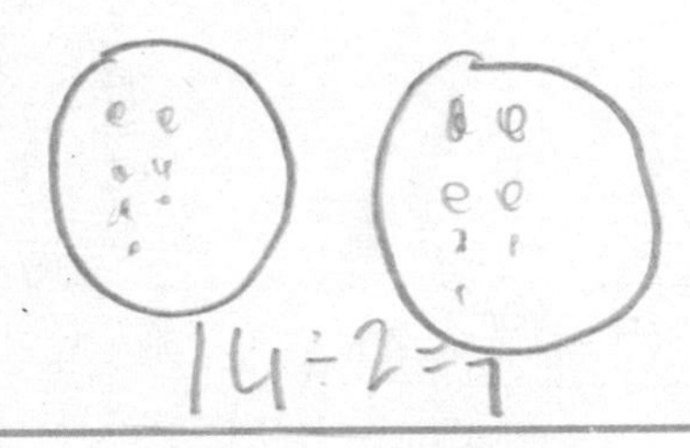	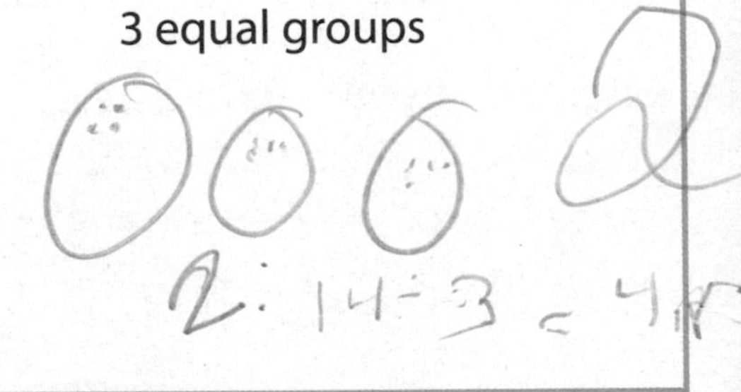
4 equal groups	**5 equal groups**
	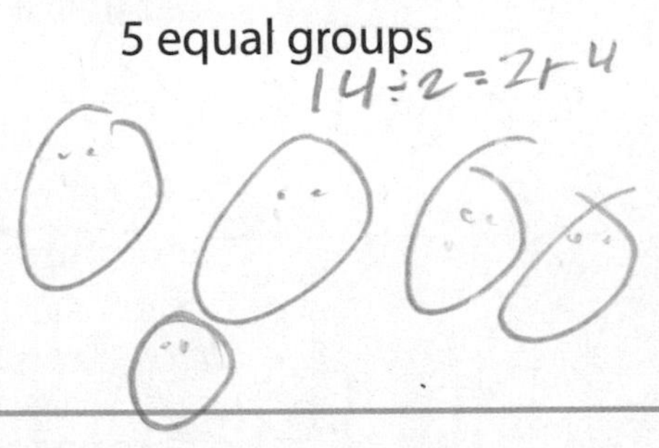

Structure Explain why the units that are left over cannot be put into a group.

Think and Grow: Find and Interpret Remainders

Sometimes you cannot divide a number evenly and there is an amount left over.

The amount left over is called the **remainder**. Use an R to represent the remainder.

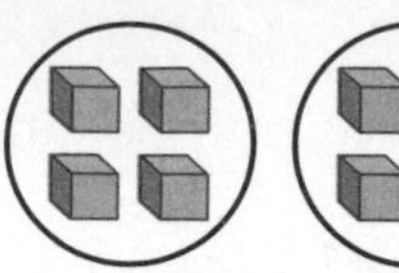 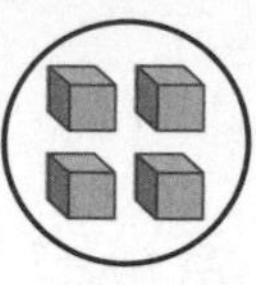 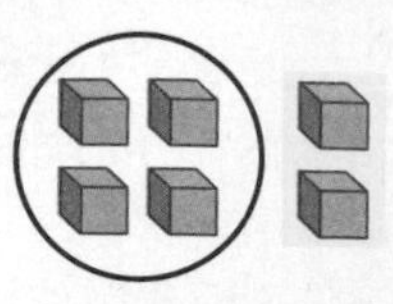

$14 \div 3 = 4$ with 2 left over

$14 \div 3 = 4 \text{ R}2$

$$3\overline{)14}\quad 4 \text{ R}2$$

Example Find $27 \div 4$.

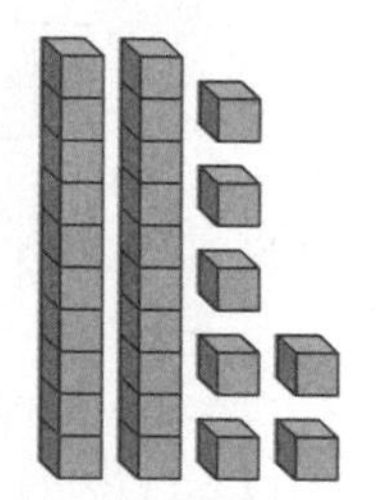

Divide 27 into 4 equal groups.

You need to regroup 2 tens as 20 ones.

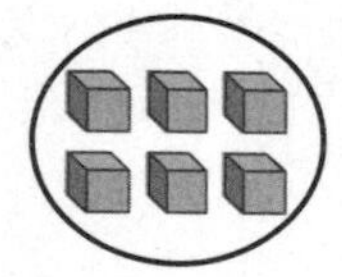 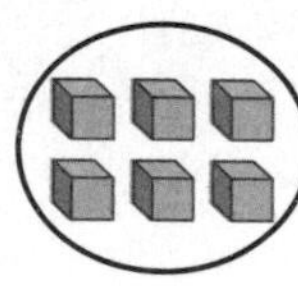 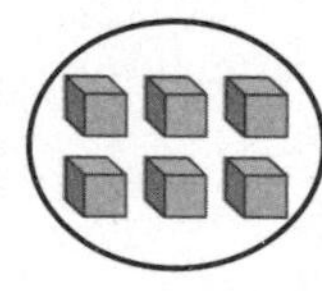 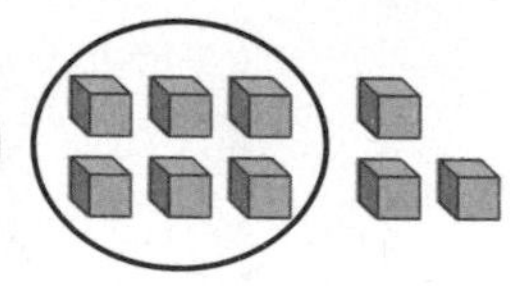

Think: $27 \div 4 = 6 \text{ R}3$ because $4 \times 6 + 3 = 27$.

Number of units in each group: ______

Number of units left over: ______

So, $27 \div 4 =$ ______ R ______.

Show and Grow *I can do it!*

Use a model to find the quotient and the remainder.

1. $19 \div 6 =$ 3 R 1

2. $34 \div 5 =$ ______ R ______

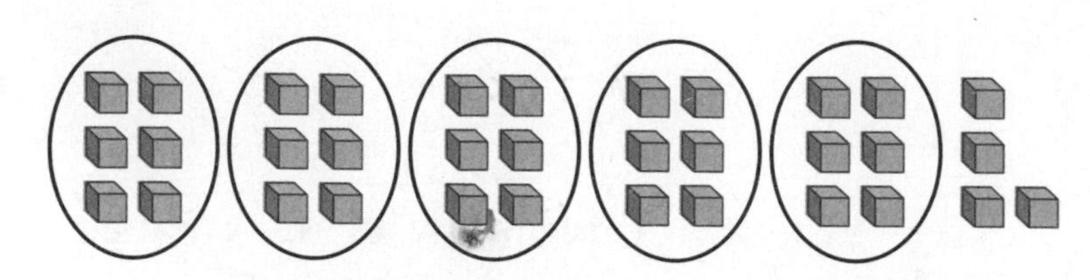

3. $26 \div 3 =$ 8 R 2

4. $20 \div 7 =$ 2 R 6

Name ____________________

Apply and Grow: Practice

Use a model to find the quotient and the remainder.

5. $13 \div 2 =$ ____ R ____

6. $25 \div 9 =$ ____ R ____

7. $28 \div 8 =$ ____ R ____

8. $15 \div 4 =$ ____ R ____

9. $29 \div 6 =$ ____ R ____

10. $11 \div 5 =$ ____ R ____

11. Descartes has 23 cat treats to divide equally among 4 friends. How many treats does he give each friend? How many treats are left over?

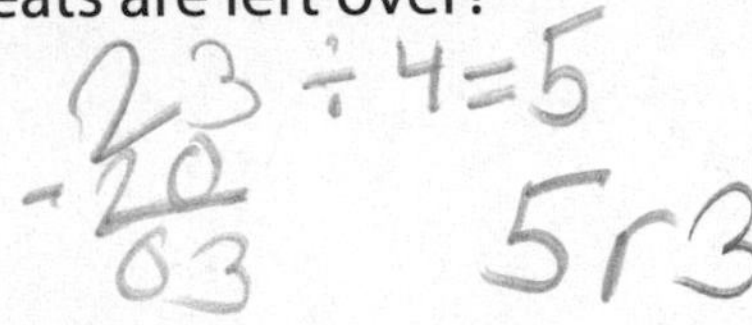

12. You have 26 markers. How many groups of 3 markers can you make? How many markers are left over?

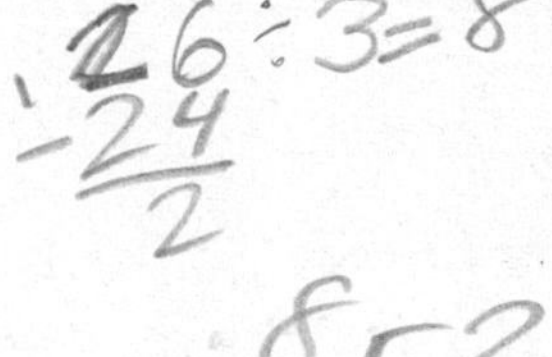

13. MP **Structure** Write a division equation represented by the model.

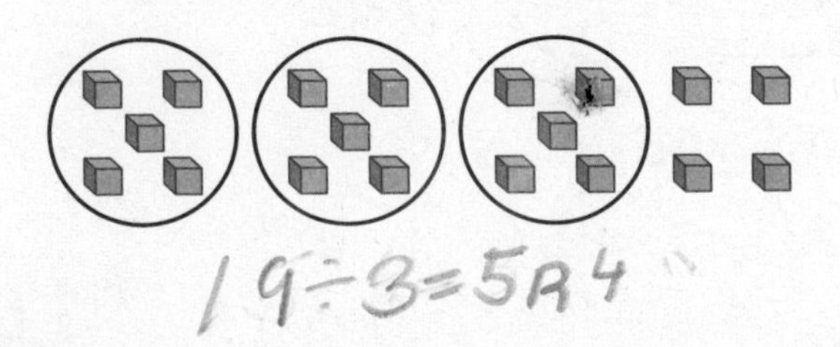

14. **YOU BE THE TEACHER** Is Newton correct? Draw a model to support your answer.

Think and Grow: Modeling Real Life

Example A water taxi transports passengers to an island. The taxi holds no more than 8 passengers at a time. There are 53 people in line to ride the water taxi.

- How many trips to the island are full?
- How many trips to the island are needed?
- How many passengers are on the last trip?

Use a model to find $53 \div 8$.

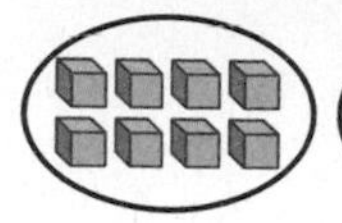 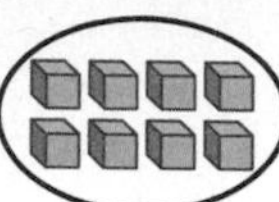 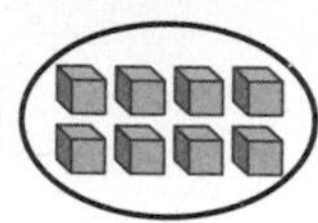 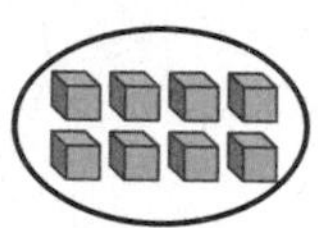 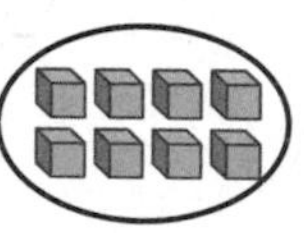 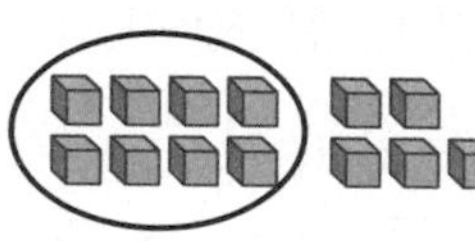

The quotient is ______. The remainder is ______.

Interpret the quotient and the remainder.

How many trips to the island are full?

The __________ is the number of trips that have 8 passengers.

So, ______ trips to the island are full.

How many trips to the island are needed?

______ trips are full and ______ trip is *not* full.

So, ______ trips are needed.

How many passengers are on the last trip?

The __________ is the number of passengers that are on the last trip.

So, ______ passengers are on the last trip.

Show and Grow I can think deeper!

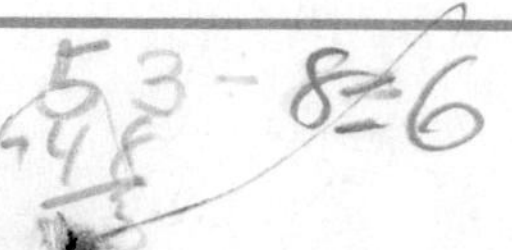

15. Tours of a crayon factory have no more than 9 guests. There are 87 guests in line to tour the factory.

- How many tours are full?
- How many tours are needed?
- How many guests are in the last tour?

Name ______________________________

Use Partial Quotients 5.4

Learning Target: Use partial quotients to divide.

Success Criteria:

- I can explain how to use an area model to divide.
- I can write partial quotients for a division problem.
- I can add the partial quotients to find a quotient.

Explore and Grow

Use the area models to find 3×12 and $36 \div 3$.

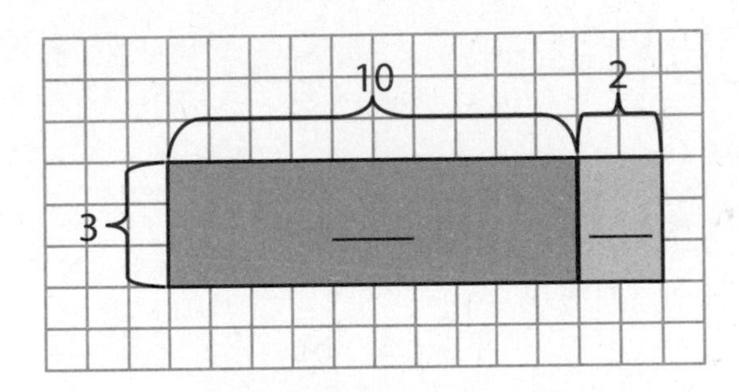

$3 \times 12 =$ _____

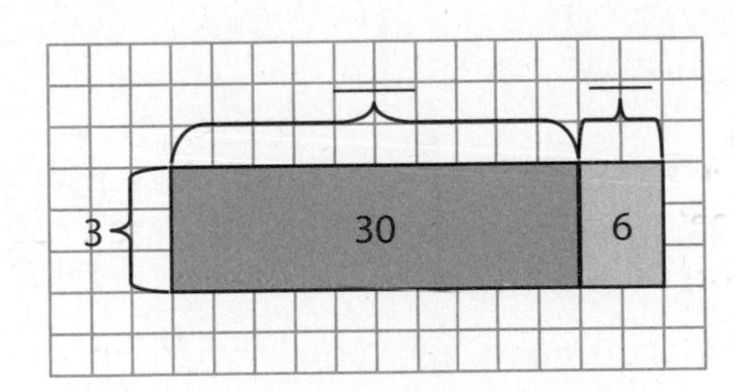

$36 \div 3 =$ _____

Reasoning How does the Distributive Property relate to each of the area models? Explain.

Think and Grow: Use Partial Quotients to Divide

To divide using **partial quotients**, subtract a multiple of the divisor that is less than the dividend. Continue to subtract multiples until the remainder is less than the divisor. The factors that are multiplied by the divisor are called partial quotients. Their sum is the quotient.

Example Use an area model and partial quotients to find 235 ÷ 5.

One Way:

	Partial Quotients
5)235	
− 100 = 5 × 20	20
135	
− 100 = 5 × 20	20
35	
− 35 = 5 × 7	+ 7
0	☐

So, 235 ÷ 5 = ______.

	20	20	7
5	100	100	35

Area = 235 square units

Another Way:

	Partial Quotients
5)235	
− 200 = 5 × 40	40
35	
− 35 = 5 × 7	+ 7
0	☐

So, 235 ÷ 5 = ______.

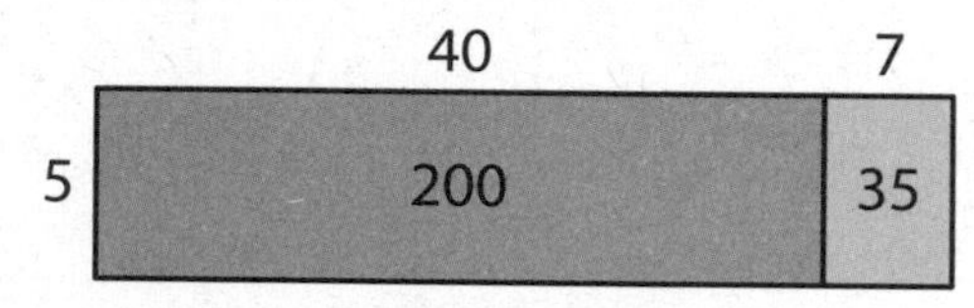

	40	7
5	200	35

Area = 235 square units

Show and Grow *I can do it!*

Use an area model and partial quotients to divide.

1. 60 ÷ 4 = ______

4)60	
− 40 = 4 × ______	☐
☐	
− 20 = 4 × ______	+ ☐
☐	☐

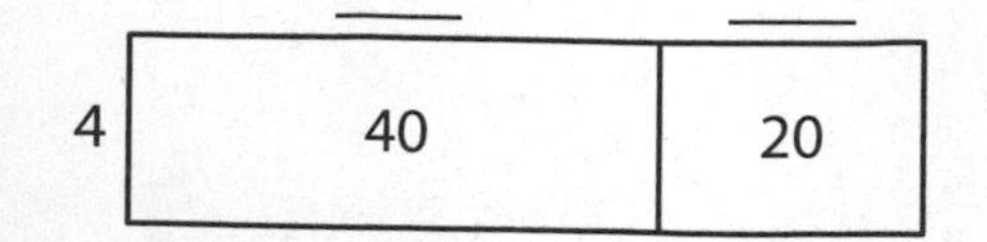

	____	____
4	40	20

2. 192 ÷ 3 = ______

3)192	
− ☐ = 3 × ______	☐
☐	
− ☐ = 3 × ______	+ ☐
☐	☐

	____	____
3	180	____

Name ____________________

Apply and Grow: Practice

3. Use an area model and partial quotients to find $264 \div 8$.

$8\overline{)264}$
$- \square = 8 \times$ ____ $\square$
$\square$
$- \square = 8 \times$ ____ $+ \square$
$\square$

8 | 240 | ____

Use partial quotients to divide.

4. $4\overline{)96}$

5. $9\overline{)405}$

6. $6\overline{)378}$

7. $7\overline{)84}$

8. $5\overline{)735}$

9. MP **Structure** Find the missing numbers.

$4\overline{)332}$
$-\ 240 = 4 \times$ ____ $\square$
92
$-\ 80 = 4 \times$ ____ $\square$
12
$-\ 12 = 4 \times$ ____ $+ \square$
0 $\square$

Think and Grow: Modeling Real Life

Example There are 8 students on each tug-of-war team. How many tug-of-war teams are there?

Use an area model and partial quotients to find 128 ÷ 8.

Field Day Activity Sign-Ups	
Activity	**Number of Students**
Kickball	107
Relay race	90
Tug-of-war	128
Volleyball	96
Water balloon toss	156

$$\begin{array}{r} 8\overline{)128} \\ -\ 80 = 8 \times 10 \\ \hline 48 \\ -\ 48 = 8 \times 6 \\ \hline 0 \end{array} \qquad \begin{array}{r} \square \\ +\ \square \\ \hline \square \end{array}$$

	10	6
8	80	48

128 ÷ 8 = ______

So, there are ______ tug-of-war teams.

Show and Grow *I can think deeper!*

Use the table above.

10. There are 5 students on each relay race team. How many relay race teams are there?

11. **DIG DEEPER!** There are 6 students on each volleyball team. There are 4 fewer students on each water balloon toss team than each volleyball team. How many of each team are there?

12. Twenty-eight students were absent on the day of sign-ups. They all decide to play kickball. There are 9 students on each kickball team. How many kickball teams are there?

Name ______________________

Homework & Practice 5.4

Learning Target: Use partial quotients to divide.

Example Use an area model and partial quotients to find $84 \div 6$.

Subtract a multiple of 6 that is less than 84. Continue to subtract multiples until the remainder is less than 6. The factors that are multiplied by 6 are called partial quotients. Their sum is the quotient.

Partial Quotients

$$\begin{array}{rll} 6\overline{)84} & & \\ -\ 60 & = 6 \times 10 & 10 \\ \hline 24 & & \\ -\ 24 & = 6 \times 4 & +\ 4 \\ \hline 0 & & \boxed{14} \end{array}$$

	10	4
6	60	24

Area = 84 square units

So, $84 \div 6 =$ __14__.

1. Use an area model and partial quotients to find $345 \div 5$.

$$\begin{array}{rll} 5\overline{)345} & & \\ -\ \square & = 5 \times \_\_\_ & \square \\ \hline \square & & \\ -\ \square & = 5 \times \_\_\_ & +\ \square \\ \hline \square & & \square \end{array}$$

	___	___
5	___	___

Use partial quotients to divide.

2. $6\overline{)90}$

3. $3\overline{)48}$

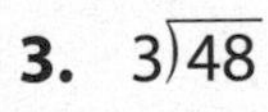

4. $8\overline{)200}$

Use partial quotients to divide.

5. $4\overline{)56}$

6. $7\overline{)511}$

7. $9\overline{)423}$

8. **YOU BE THE TEACHER** Descartes finds $952 \div 8$. Is he correct? Explain.

$$
\begin{array}{rll}
8\overline{)952} & & \\
-\ 800 & = 8 \times 100 & 100 \\
\hline
152 & & \\
-\ 80 & = 8 \times 10 & 10 \\
\hline
72 & & \\
-\ 72 & = 8 \times 9 & +\ 9 \\
\hline
0 & & 119
\end{array}
$$

9. **Writing** Explain how you can solve a division problem more than one way using partial quotients.

10. **Modeling Real Life** Each shelter animal gets 3 toys. How many shelter animals are there?

Animal Shelter Donations	
Item	**Number Donated**
Bedding	96
Toys	168
Bowls	120
Leashes	72
Bags of food	231

Review & Refresh

Find the product.

11. $40 \times 70 =$ ______

12. $30 \times 58 =$ ______

13. $62 \times 90 =$ ______

Name ___________________________

Use Partial Quotients with a Remainder

5.5

Learning Target: Use partial quotients to divide and find remainders.

Success Criteria:
- I can use partial quotients to divide.
- I can find a remainder.

Use an area model to find 125 ÷ 5.

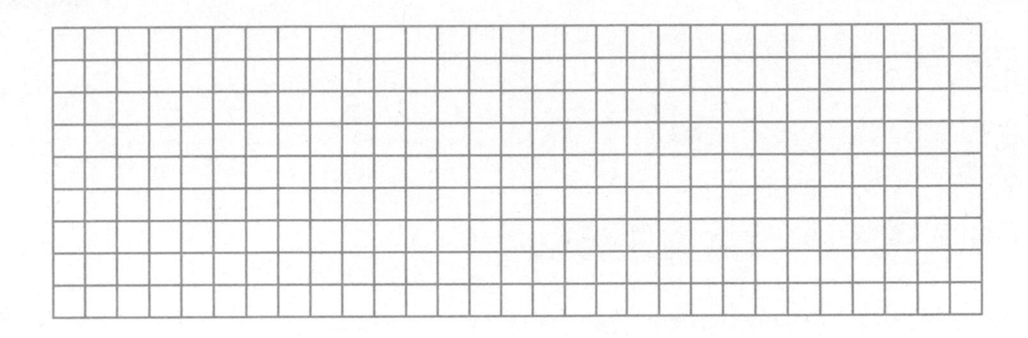

Can you use an area model to find 128 ÷ 5? Explain your reasoning.

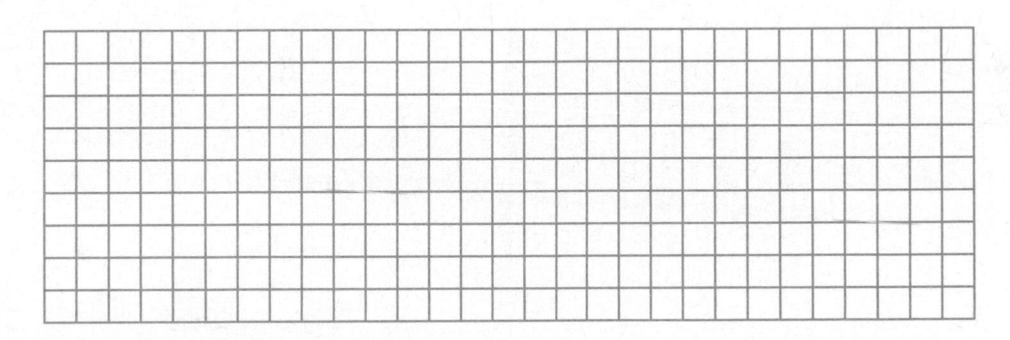

MP **Construct Arguments** Explain to your partner how your model shows that 5 does *not* divide evenly into 128.

Think and Grow: Use Partial Quotients with a Remainder

Example Use partial quotients to find 2,918 ÷ 4.

$$
\begin{array}{r l}
4\overline{)2{,}918} & \\
-\ 2{,}800 = 4 \times 700 & \square \\
118 & \\
-\ 100 = 4 \times 25 & \square \\
18 & \\
-\ 16 = 4 \times 4 & + \square \\
2 & \square \text{ R } \square
\end{array}
$$

So, 2,918 ÷ 4 = ______ R ______.

Show and Grow *I can do it!*

Use partial quotients to divide.

1. 82 ÷ 3 = ______

$$
\begin{array}{r l}
3\overline{)82} & \\
-\ 60 = 3 \times \underline{\quad} & \square \\
\square & \\
-\ 21 = 3 \times \underline{\quad} & + \square \\
\square & \square \text{ R } \square
\end{array}
$$

2. 754 ÷ 9 = ______

$$
\begin{array}{r l}
9\overline{)754} & \\
-\ 720 = 9 \times \underline{\quad} & \square \\
\square & \\
-\ 27 = 9 \times \underline{\quad} & + \square \\
\square & \square \text{ R } \square
\end{array}
$$

3. $8\overline{)460}$

4. $5\overline{)3{,}242}$

5. $6\overline{)5{,}850}$

Name ______________________________

Apply and Grow: Practice

Use partial quotients to divide.

6. $5\overline{)63}$

7. $7\overline{)401}$

8. $4\overline{)5{,}237}$

9. $9\overline{)256}$

10. $8\overline{)945}$

11. $2\overline{)7{,}043}$

12. The third, fourth, and fifth grades make 146 science projects for a fair. Did each grade make the same number of projects? Explain.

13. MP **Structure** Newton found $315 \div 6$. Explain how the steps would be different if he had used 50 as the first partial quotient?

$$
\begin{array}{rll}
6\overline{)315} & & \\
-\ 240 & = 6 \times 40 & 40 \\
\hline
75 & & \\
-\ 60 & = 6 \times 10 & 10 \\
\hline
15 & & \\
-\ 12 & = 6 \times 2 & +\ 2 \\
\hline
3 & & 52\text{ R}3
\end{array}
$$

So, $315 \div 6 = 52$ R3.

Think and Grow: Modeling Real Life

Example There are 1,862 people attending a mud run. Each wave of runners can have 8 people. How many waves of runners are needed?

Use partial quotients to find $1,862 \div 8$.

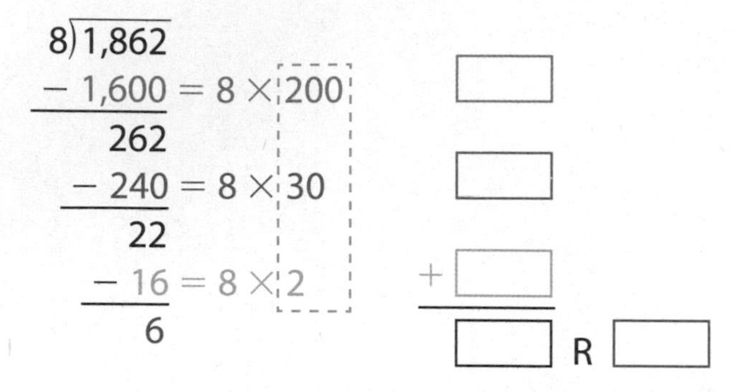

$1,862 \div 8 =$ ______ R ______

Interpret the quotient and the remainder.

The quotient is ______. This means that ______ waves of runners will have 8 people.

The remainder is ______. This means that 1 wave of runners will have ______ people.

So, ______ waves of runners are are needed.

Show and Grow *I can think deeper!*

14. A juice factory has 768 fluid ounces of juice for guests to sample. A worker pours the juice into 5-fluid ounce cups. How many cups does the worker fill?

15. A toy company designs 214 collectible figures. The company releases 6 of the figures each month. How many months will it take the company to release all of the collectible figures? How many years will it take?

Name ________________________________

Homework & Practice 5.5

Learning Target: Use partial quotients to divide and find remainders.

Example Use partial quotients to find 230 ÷ 9.

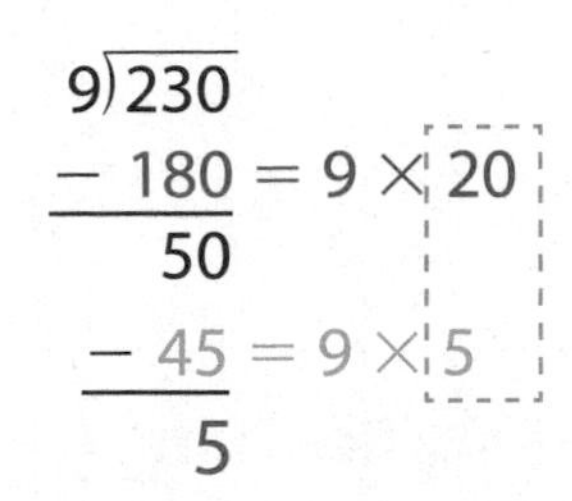

20
+ 5
25 R 5

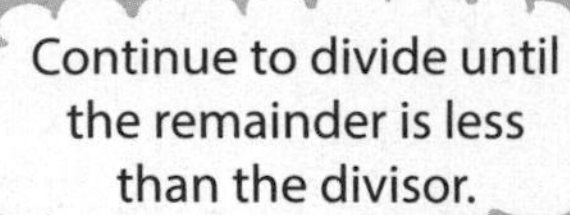

So, 230 ÷ 9 = __25__ R __5__.

Use partial quotients to divide.

1. 4)67

2. 2)715

3. 5)1,308

4. 3)516

5. 9)2,497

6. 6)831

Use partial quotients to divide.

7. $9\overline{)476}$

8. $7\overline{)2,254}$

9. $4\overline{)3,018}$

10. MP **Reasoning** Show how to use the least number of partial quotients to find 3,526 ÷ 4.

11. **Modeling Real Life** A gardening center has 1,582 pots to fill. Each bag of soil can fill 4 pots. How many bags of soil are needed?

12. DIG DEEPER! You have 178 photos. You put 3 photos on each page of an album. Your friend has 354 photos. She puts 6 photos on each page of an album. Who uses more pages? Explain.

Review & Refresh

13. An Olympic swimmer wants to eat 10,000 calories each day. He eats 3,142 calories at breakfast and 3,269 calories at lunch. How many more calories must the swimmer eat to reach his goal?

Name ______________________

Divide Two-Digit Numbers by One-Digit Numbers

5.6

Learning Target: Divide two-digit numbers by one-digit numbers.

Success Criteria:

- I can divide to find the partial quotients.
- I can show how to regroup 1 or more tens.
- I can use place value to record the partial quotients.

Use a model to find each quotient. Draw each model.

84 ÷ 4

85 ÷ 5

Construct Arguments Explain to your partner how your methods for finding the quotients above are the same. Then explain how they are different.

Think and Grow: Use Regrouping to Divide

Example Find $79 \div 3$. Think: 79 is 7 tens and 9 ones.

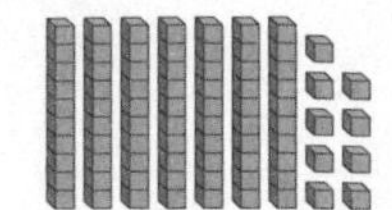

Divide the tens.

$$\begin{array}{r} 2 \\ 3\overline{)79} \\ -\,6 \\ \hline 1 \end{array}$$

Divide: 7 tens ÷ 3

Multiply: 2 tens × 3

Subtract: 7 tens − 6 tens

There is 1 ten left over.

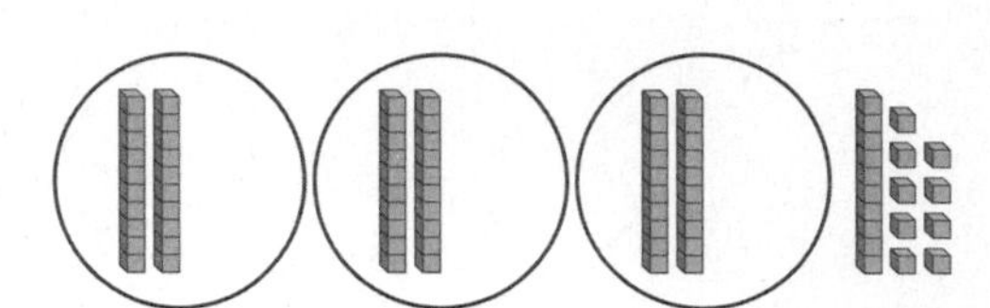

Regroup.

$$\begin{array}{r} 2 \\ 3\overline{)79} \\ -\,6\downarrow \\ \hline 19 \end{array}$$

Regroup 1 ten as 10 ones.

10 ones + 9 ones = 19 ones

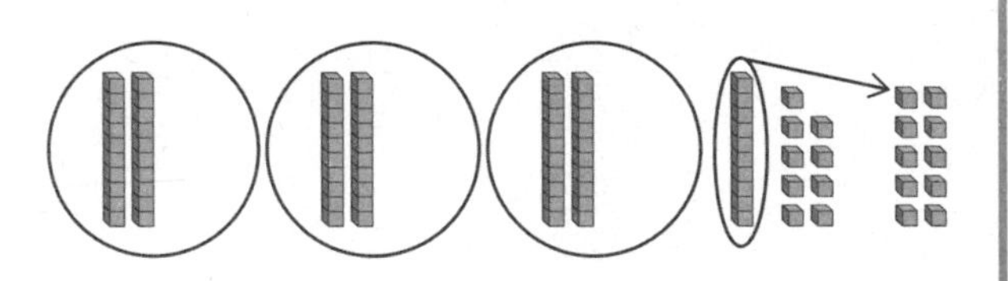

Divide the ones.

$$\begin{array}{r} 26 \text{ R}1 \\ 3\overline{)79}\phantom{\text{ R}1} \\ -\,6\downarrow\phantom{\text{ R}1} \\ \hline 19\phantom{\text{ R}1} \\ -\,18\phantom{\text{ R}1} \\ \hline 1\phantom{\text{ R}1} \end{array}$$

Divide: 19 ones ÷ 3

Multiply: 6 ones × 3

Subtract: 19 ones − 18 ones

There is 1 one left over.

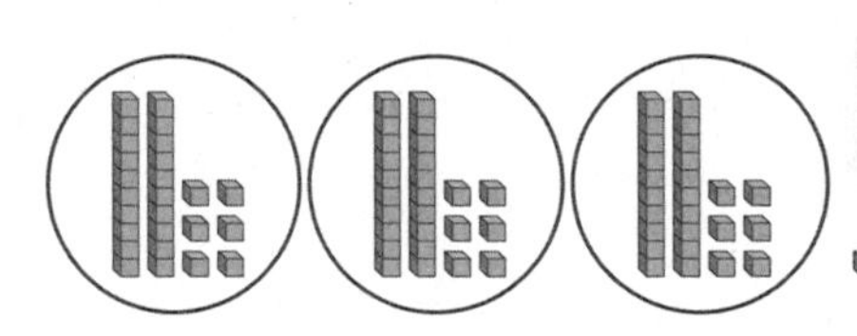

Check:
$26 \times 3 + 1 = 78 + 1 = 79$

So, $79 \div 3 =$ ______ R ______.

Show and Grow I can do it!

Divide. Then check your answer.

1. $6\overline{)96}$ with blanks: □, − □ ↓, □, − □, □

2. $2\overline{)88}$ □

3. $5\overline{)74}$ □ R ______

Name ____________________

Apply and Grow: Practice

Divide. Then check your answer.

4. $5\overline{)60}$

5. $6\overline{)70}$

6. $8\overline{)90}$

7. $3\overline{)93}$

8. $2\overline{)45}$

9. $3\overline{)64}$

10. $6\overline{)42}$

11. $8\overline{)36}$

12. $7\overline{)50}$

13. Writing Explain how you can use estimation to check the reasonableness of your answer when dividing a two-digit number by a one-digit number.

14. MP Structure Find the missing numbers.

```
    1□ R□
 5)7□
 − □↓
   □2
 − 2 0
    □
```

Think and Grow: Modeling Real Life

Example A house cat has 64 muscles in its ears. It has the same number of muscles in each ear. How many muscles does the house cat have in each ear?

The house cat has 2 ears, so find $64 \div 2$.

Think: 64 is 6 tens and 4 ones.

Divide the tens.

$$\begin{array}{r} 3 \\ 2\overline{)64} \\ -\,6 \\ \hline 0 \end{array}$$

Divide: 6 tens ÷ 2

Multiply: 3 tens × 2

Subtract: 6 tens − 6 tens

There are 0 tens left over.

Because there are no tens left over, there are no tens to regroup. So, divide the ones.

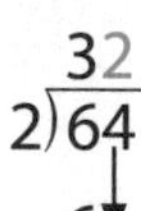

$$\begin{array}{r} 32 \\ 2\overline{)64} \\ -\,6 \\ \hline 04 \\ -4 \\ \hline 0 \end{array}$$

Divide: 4 ones ÷ 2

Multiply: 2 ones × 2

Subtract: 4 ones − 4 ones

There are 0 ones left over.

$64 \div 2 =$ ______ So, the house cat has ______ muscles in each ear.

Show and Grow I can think deeper!

15. You earn 5 cents for each plastic bottle you recycle. You recycle some bottles and earn 75 cents. How many bottles did you recycle?

16. DIG DEEPER! A cross-country runner must run 80 miles in 1 week. He wants to run about the same number of miles each day. How many miles should he run each day? How can you interpret the remainder?

17. Admission to a go-kart park costs a total of $78 for 3 adults and 3 children. The price is the same for all ages. What is the cost of admission for each person?

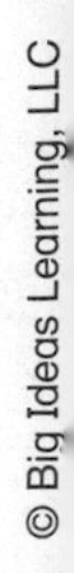

Name ______________________________

5.7 Divide Multi-Digit Numbers by One-Digit Numbers

Learning Target: Divide multi-digit numbers by one-digit numbers.

Success Criteria:

- I can use place value to divide.
- I can show how to regroup thousands, hundreds, or tens.
- I can find a quotient and a remainder.

Explore and Grow

Use a model to divide.
Draw each model.

$348 \div 3$

$148 \div 3$

Reasoning Explain why the quotient of $148 \div 3$ does *not* have a digit in the hundreds place.

Think and Grow: Use Regrouping to Divide

Example Find 907 ÷ 5.

Estimate: 1,000 ÷ 5 = ______

Divide the hundreds.

$$\begin{array}{r} 1 \\ 5\overline{)907} \\ -5 \\ \hline 4 \end{array}$$

9 hundreds ÷ 5

1 hundred × 5

9 hundreds − 5 hundreds

There are 4 hundreds left over.

Divide the tens.

$$\begin{array}{r} 18 \\ 5\overline{)907} \\ -5 \\ \hline 40 \\ -40 \\ \hline 0 \end{array}$$

Regroup 4 hundreds as 40 tens: 40 ÷ 5

8 tens × 5

40 tens − 40 tens

There are 0 tens left over.

Divide the ones.

$$\begin{array}{r} 181 \text{ R}2 \\ 5\overline{)907} \\ -5 \\ \hline 40 \\ -40 \\ \hline 07 \\ -5 \\ \hline 2 \end{array}$$

There are no tens to regroup, so divide the ones: 7 ones ÷ 5

1 one × 5

7 ones − 5 ones

There are 2 ones left over.

So, 907 ÷ 5 = ______ R ______.

Check: Because ______ R ______ is close to the estimate, the answer is reasonable.

Show and Grow *I can do it!*

Divide. Then check your answer.

1. $4\overline{)531}$ = ☐ R ______

2. $5\overline{)7{,}180}$ = ☐

3. $7\overline{)8{,}385}$ = ☐ R ______

Name ______________________________

Apply and Grow: Practice

Divide. Then check your answer.

4. $5\overline{)6,381}$

5. $3\overline{)4,605}$

6. $6\overline{)820}$

7. $6\overline{)7,039}$

8. $4\overline{)855}$

9. $2\overline{)367}$

10. $8\overline{)9,692}$

11. $7\overline{)8,345}$

12. $7\overline{)971}$

13. There are 8,274 people at an air show. The people are divided equally into 6 sections. How many people are in each section?

14. **YOU BE THE TEACHER** Newton finds 120 ÷ 5. Is he correct? Explain.

$$\begin{array}{r} 114 \\ 5\overline{)120} \\ -\ 5 \\ \hline 7 \\ -\ 5 \\ \hline 20 \\ -\ 20 \\ \hline 0 \end{array}$$

Think and Grow: Modeling Real Life

Example There are 1,014 toy car tires at a factory. Each car needs 4 tires. How many toy cars can the factory workers make with the tires?

Each car needs 4 tires, so find 1,014 ÷ 4.

1 thousand *cannot* be shared among 4 groups without regrouping. So, regroup 1 thousand as 10 hundreds.

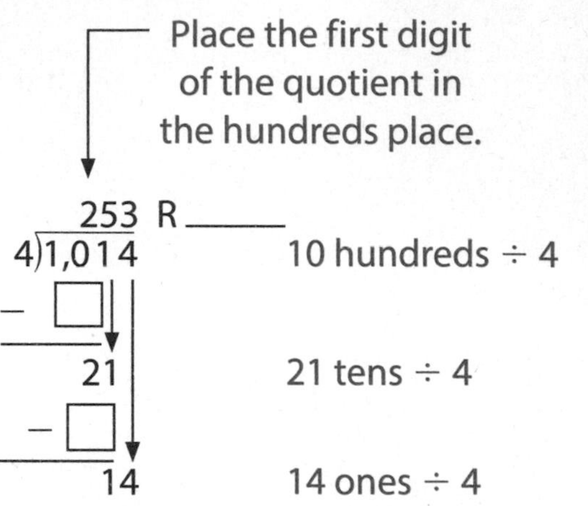

1,014 ÷ 4 = ______ R ______

Interpret the quotient and the remainder.

The quotient is ______. The factory workers can make ______ toy cars.

The remainder is ______. There are ______ tires left over.

Show and Grow *I can think deeper!*

15. A principal orders 750 tablets. The distributor can fit 8 tablets in each box. How many boxes are needed to ship all of the tablets?

16. An athlete's heart rate after a 5-mile run is 171 beats per minute, which is 3 times as fast as her resting heart rate. What is the athlete's resting heart rate?

17. A car costs $5,749. The taxes and fees for the car cost an additional $496. A customer uses a 5-year interest-free loan to buy the car. How much money will the customer pay for the car each year?

Name ______________________________

Divide by One-Digit Numbers 5.8

Learning Target: Divide by one-digit numbers.

Success Criteria:

- I can use place value to divide.
- I can explain why there might be a 0 in the quotient.
- I can find a quotient and a remainder.

Explore and Grow

Use a model to find each quotient.
Draw each model.

$312 \div 3$

$312 \div 4$

Structure Compare your models for each quotient. What is the same? What is different? What do you think this means when using regrouping to divide?

Think and Grow: Divide by One-Digit Numbers

Example Find 4,829 ÷ 8.

4 thousands *cannot* be shared among 8 groups without regrouping.
So, regroup 4 thousands as 40 hundreds and combine with 8 hundreds.

Divide the hundreds.

```
     6
 8)4,829
  − 48
     0
```

48 hundreds ÷ 8
6 hundreds × 8
48 hundreds − 48 hundreds
There are 0 hundreds left over.

Divide the tens.

```
     60
 8)4,829
  − 48↓
     02
  −   0
      2
```

2 tens *cannot* be shared among 8 groups without regrouping. So, place a zero in the quotient.

0 tens × 8
2 tens − 0 tens
There are 2 tens left over.

When a place value in the dividend cannot be divided by the divisor without regrouping, write a zero in the quotient.

Divide the ones.

```
     603 R5
 8)4,829
  − 48
     02
   −  0
     29
   − 24
      5
```

Regroup 2 tens as 20 ones and combine with 9 ones.

29 ones ÷ 8
3 ones × 8
29 ones − 24 ones
There are 5 ones left over.

So, 4,829 ÷ 8 = ______ R ______.

Show and Grow I can do it!

Divide. Then check your answer.

1. $7\overline{)756}$ ☐

2. $6\overline{)364}$ ☐ R ______

3. $3\overline{)3,190}$ ☐ R ______

Name

Apply and Grow: Practice

Divide. Then check your answer.

4. $2\overline{)81}$

5. $4\overline{)428}$

6. $6\overline{)842}$

7. $3\overline{)2{,}724}$

8. $9\overline{)635}$

9. $6\overline{)1{,}442}$

10. $6\overline{)303}$

11. $5\overline{)2{,}530}$

12. $8\overline{)7{,}209}$

13. The 5 developers of a phone app earn a profit of $4,535 this month. They divide the profit equally. How much money does each developer get?

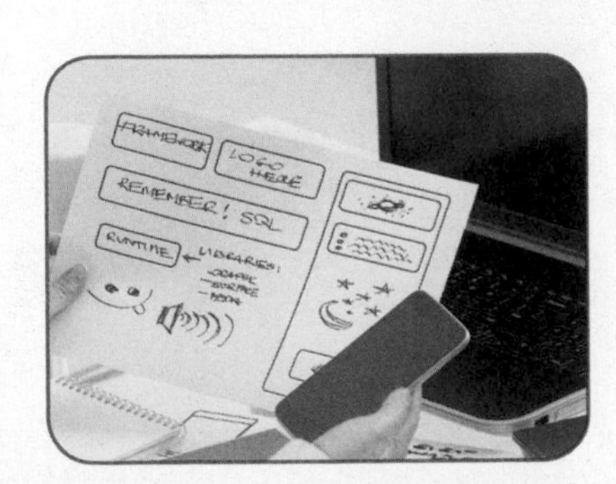

14. **YOU BE THE TEACHER** Newton finds 817 ÷ 4. Is he correct? Explain.

```
   24  R1
 4)817
 - 8
    17
  - 16
     1
```

Think and Grow: Modeling Real Life

Example Seven players are placed on each basketball team. Remaining basketball players are added to the teams, so some of the teams have 8 players. How many basketball teams have 7 players? 8 players?

Sport	Number of Players
Soccer	2,476
Basketball	1,839
Flag football	3,214
Ball hockey	952

There are 1,839 players signed up for basketball, so find 1,839 ÷ 7.

1 thousand *cannot* be shared among 7 groups without regrouping. So, regroup 1 thousand as 10 hundreds and combine with 8 hundreds.

```
    ____ R____
 7)1,839        18 hundreds ÷ 7
  − □
    43          43 tens ÷ 7
   − □
     19         19 ones ÷ 7
    − □
      □         1,839 ÷ 7 = ____ R ____
```


Interpret the quotient and the remainder.

The quotient is ______. So, there are ______ basketball teams in all.

The remainder is ______. So, ______ basketball teams have 8 players.

Subtract to find how many teams have 7 players. ______ − ______ = ______

So, ______ basketball teams have 7 players and ______ have 8 players.

Show and Grow *I can think deeper!*

Use the table above.

15. Nine players are placed on each ball hockey team. Remaining players are added to the teams, so some of the teams have 10 players. How many ball hockey teams have 9 players? 10 players?

16. Eighty-four players who signed up to play soccer decide not to play. Eight players are placed on each soccer team. How many soccer teams are there?

Name ______________________

Homework & Practice 5.8

Learning Target: Divide by one-digit numbers.

Example Find 283 ÷ 4.

2 hundreds *cannot* be shared among 4 groups without regrouping. So, regroup 2 hundreds as 20 tens and combine with 8 tens.

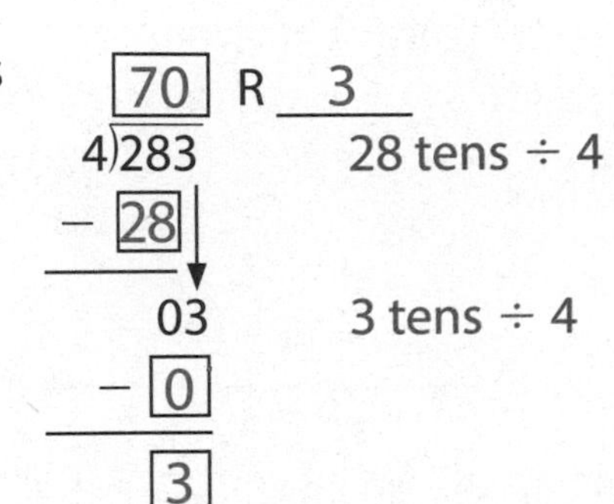

When a place value in the dividend cannot be divided by the divisor without regrouping, write a zero in the quotient.

So, 283 ÷ 4 = 70 R 3.

Divide. Then check your answer.

1. 8)832

2. 7)215 R ____

3. 5)5,078 R ____

4. 7)94

5. 6)731

6. 4)6,514

7. 3)62

8. 5)548

9. 2)4,136

Divide. Then check your answer.

10. $7\overline{)214}$

11. $4\overline{)321}$

12. $6\overline{)5,162}$

13. $2\overline{)7,301}$

14. $5\overline{)603}$

15. $3\overline{)6,082}$

16. There are 450 pounds of grapes for a grape stomping contest. They are divided equally into 5 barrels. How many pounds of grapes are in each barrel?

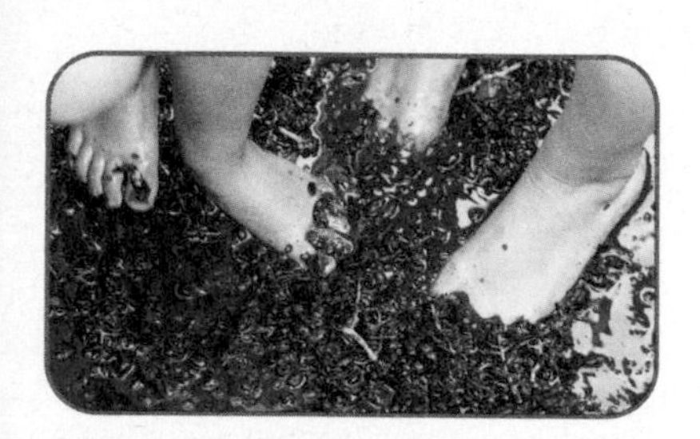

17. **DIG DEEPER!** How could you change the dividend in Exercise 11 so that there would be no remainder? Explain.

18. **Modeling Real Life** Five actresses are placed on each team. Remaining actresses are added to the teams, so some of the teams have 6 actresses. How many teams have 5 actresses? 6 actresses?

Celebrity Game Show Sign-Ups

Celebrity	Number of Players
Singers	105
Musicians	81
Actors	197
Actresses	202

Review & Refresh

Write the value of the underlined digit.

19. 8$\underline{6}$,109

20. $\underline{1}$5,327

21. $\underline{9}$14,263

22. 284,5$\underline{0}$5

Name ______________________________

Problem Solving: Division 5.9

Learning Target: Solve multi-step word problems involving division.

Success Criteria:

- I can understand a problem.
- I can make a plan to solve using letters to represent the unknown numbers.
- I can solve a problem using an equation.

Make a plan to solve the problem.

A fruit vendor has 352 green apples and 424 red apples. The vendor uses all of the apples to make fruit baskets. He puts 8 apples in each basket. How many fruit baskets does the vendor make?

Make Sense of Problems The vendor decides that each basket should have 8 of the same colored apples. Does this change your plan to solve the problem? Will this change the answer? Explain.

Think and Grow: Problem Solving: Division

Example The speed of sound in water is 1,484 meters per second. Sound travels 112 more than 4 times as many meters per second in water as it does in air. What is the speed of sound in air?

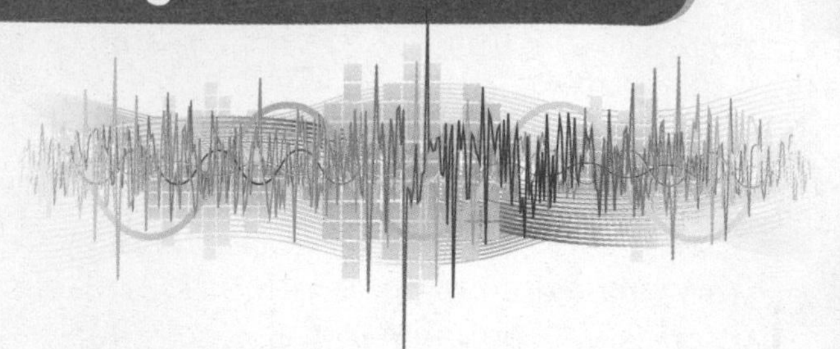

Understand the Problem

What do you know?

- The speed of sound in water is 1,484 meters per second.
- Sound travels 112 more than 4 times as many meters per second in water as it does in air.

What do you need to find?

- You need to find the speed of sound in air.

Make a Plan

How will you solve?

- Subtract 112 from 1,484 to find 4 times the speed of sound in air.
- Then divide the difference by 4 to find the speed of sound in air.

Solve

Step 1: $1{,}484 - 112 = d$

d is the unknown difference.

$$\begin{array}{r} 1{,}484 \\ -\ \ 112 \\ \hline \square \end{array}$$

$d =$ ______

Step 2: $d \div 4 = a$

a is the unknown value.

$a =$ ______

The speed of sound in air is ______ meters per second.

Show and Grow I can do it!

1. Explain how you can check whether your answer above is reasonable.

Name ______________________________

Apply and Grow: Practice

Understand the problem. What do you know? What do you need to find? Explain.

2. A surf shop owner divides 635 stickers evenly among all of her surfboards. Each surfboard has 3 tiki stickers and 2 turtle stickers. How many surfboards does she have?

3. There are 1,008 projects in a science fair. The projects are divided equally into 9 rooms. Each room has 8 equal rows of projects. How many projects are in each row?

Understand the problem. Then make a plan. How will you solve? Explain.

4. Of 78 students who work on a mural, 22 students design it, and the rest of the students paint it. The painters are divided equally among 4 areas of the mural. How many painters are assigned to each area?

5. The Winter Olympics occur twice every 8 years. How many times will the Winter Olympics occur in 200 years?

6. A party planner wants to put 12 balloons at each of 15 tables. The balloons come in packages of 8. How many packages of balloons must the party planner buy?

7. An art teacher has 8 boxes of craft sticks. Each box has 235 sticks. The students use the sticks to make as many hexagons as possible. How many sticks are *not* used?

Think and Grow: Modeling Real Life

Example A book enthusiast has $200 to buy an e-reader and e-books. He uses a $20 off coupon and buys the e-reader shown. Each e-book costs $6. How many e-books can the book enthusiast buy?

Think: What do you know? What do you need to find? How will you solve?

Step 1: How much money does the book enthusiast pay for the e-reader?

Subtract $20 from $119.

$$\begin{array}{r} 119 \\ -\ \ 20 \\ \hline \square \end{array}$$

Step 2: Subtract to find how much money he has left to spend on e-books.

$200 – ______ = *d*

d is the unknown difference.

$$\begin{array}{r} 200 \\ -\ \square \\ \hline \square \end{array}$$

d = ______

Step 3: Use *d* to find the number of e-books the book enthusiast can buy.

$d \div 6 = q$

q is the unknown value.

$6\overline{)\square}$ quotient $\square$ R ______

q = ______ R ______

So, the book enthusiast can buy ______ e-books.

Show and Grow I can think deeper!

8. You run 17 laps around a track. Newton runs 5 times as many laps as you. Descartes runs 35 more laps than Newton. Eight laps around the track are equal to 1 mile. How many miles does Descartes run?

Name ______________________________

Homework & Practice 5.9

Learning Target: Solve multi-step word problems involving division.

Example The Statue of Liberty was a gift from the people of France to the people of the United States of America. The Eiffel Tower in France is 986 feet tall. It is 234 feet shorter than 4 times the height of the Statue of Liberty. How tall is the Statue of Liberty?

Eiffel Tower

Statue of Liberty

Think: What do you know? What do you need to find? How will you solve?

Step 1: $986 + 234 = x$

x is the unknown sum.

$$\begin{array}{r} 986 \\ +\ 234 \\ \hline \boxed{1{,}220} \end{array}$$

$x =$ 1,220

Step 2: $x \div 4 = y$

y is the unknown quotient.

$$4\overline{)\boxed{1{,}220}} = \boxed{305}$$

$y =$ 305

The Statue of Liberty is 305 feet tall.

Understand the problem. Then make a plan. How will you solve? Explain.

1. You borrow a 235-page book from the library. You read 190 pages. You have 3 days left until you have to return the book. You want to read the same number of pages each day to finish the book. How many pages should you read each day?

2. There are 24 fourth graders and 38 fifth graders traveling to a math competition. If 8 students can fit into each van, how many vans are needed?

3. Your class has 3 bags of buttons to make riding horses for a relay race. Each bag has 54 buttons. What is the greatest number of horses your class can make?

Each horse needs 4 buttons.

4. Factory workers make 2,597 small, 2,597 medium, and 2,597 large plush toys. The workers pack the toys into boxes with 4 toys in each box. How many toys are left over?

5. **Writing** Write and solve a two-step word problem that can be solved using division.

6. **Modeling Real Life** You exercise for 300 minutes this week. Outside of jogging, you divide your exercising time equally among 3 other activities. How many minutes do you spend on each of your other 3 activities?

M: jog 35 min
W: jog 35 min
F: jog 35 min

7. **Modeling Real Life** Drones are used to help protect orangutans and their habitats. A drone takes a picture every 2 seconds. How many pictures does the drone take in 30 minutes?

Review & Refresh

Find the product. Check whether your answer is reasonable.

8. Estimate: ______

$41 \times 22 =$ ______

9. Estimate: ______

$87 \times 19 =$ ______

10. Estimate: ______

$36 \times 59 =$ ______

Name ___________________________

Performance Task 5

The students in fourth grade go on a field trip to a planetarium.

1. The teachers have $760 to buy all of the tickets for the teachers and students. They receive less than $6 in change.

   a. Each ticket costs $6. How many tickets do the teachers buy?

   b. Exactly how much money is left over?

   c. There are 6 groups on the field trip. Each group has 1 teacher. There are an equal number of students in each group. How many students are in each group?

   d. Two groups can be in the planetarium for each show. The planetarium has 7 rows of seats with 8 seats in each row. How many seats are empty during each show?

2. The groups will be at the planetarium from 11:00 A.M. until 2:30 P.M. During that time they will rotate through 7 events: the planetarium show, 5 activities, and lunch. The planetarium show lasts 45 minutes. Each activity lasts 22 minutes. Students have 5 minutes between each event. How long does each group have to eat lunch?

3. You learn that the distance around Mars is about twice the distance around the moon. The distance around Mars is 13,263 miles. To find the distance around the moon, do you think an estimate or an exact answer is needed? Explain.

Division Dots

Directions:

1. Players take turns connecting two dots, each using a different color.
2. On your turn, connect two dots, vertically or horizontally. If you close a square around a division problem, find and write the quotient and the remainder. If you do not close a square, your turn is over.
3. Continue playing until all division problems are solved.
4. The player with the most completed squares wins!

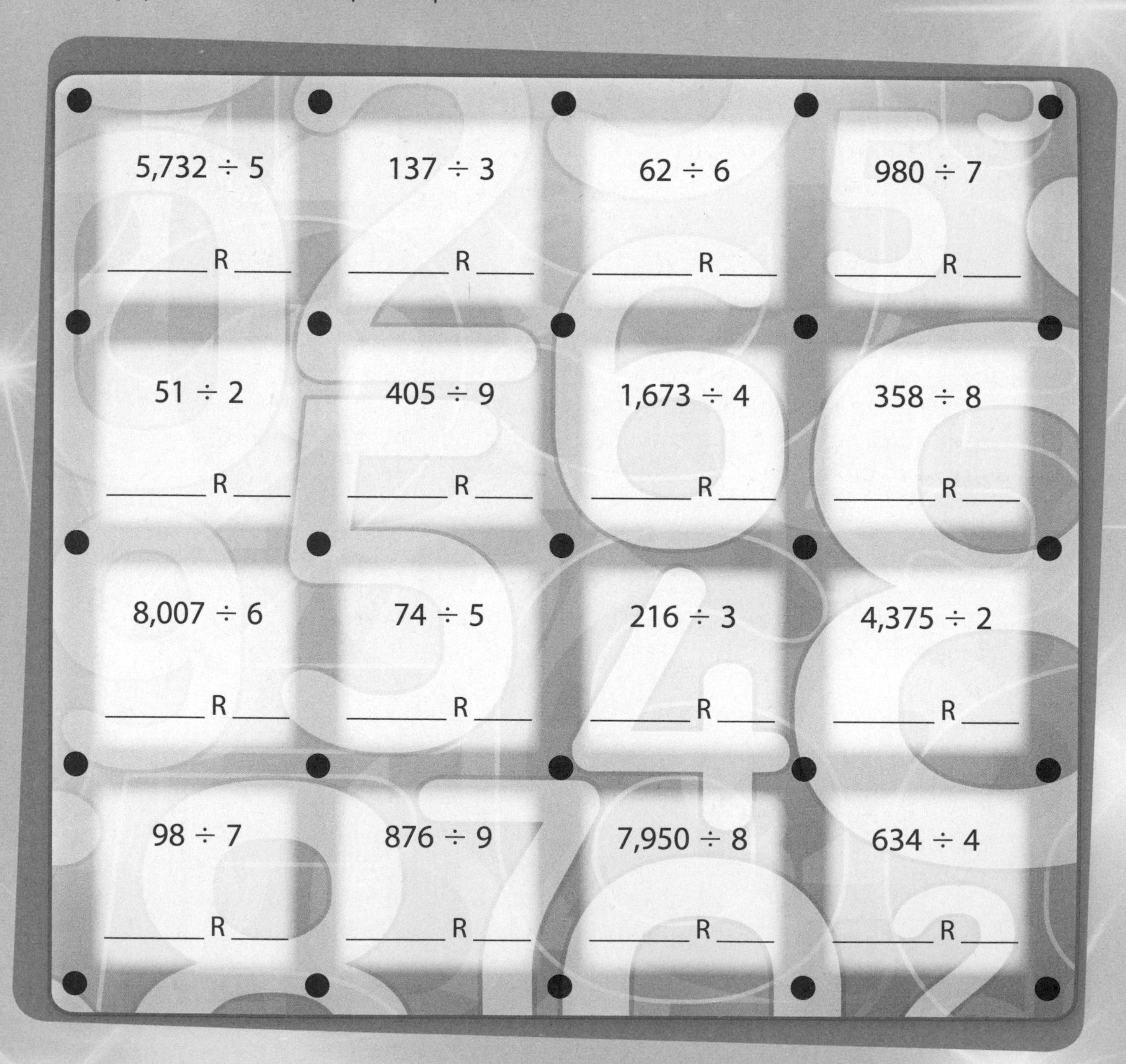

Name ______________________________

Chapter Practice

5.1 Divide Tens, Hundreds, and Thousands

Find the quotient.

1. 90 ÷ 9 = _____	**2.** 560 ÷ 7 = _____	**3.** 2,700 ÷ 9 = _____
4. 240 ÷ 4 = _____	**5.** 4,500 ÷ 5 = _____	**6.** 60 ÷ 6 = _____
7. 1,600 ÷ 8 = _____	**8.** 30 ÷ 3 = _____	**9.** 540 ÷ 9 = _____

Find the missing number.

10. 720 ÷ _____ = 80

11. _____ ÷ 7 = 10

12. 1,800 ÷ _____ = 600

5.2 Estimate Quotients

Estimate the quotient.

13. 47 ÷ 7

14. 593 ÷ 6

Find two estimates that the quotient is between.

15. 261 ÷ 8

16. 7,012 ÷ 9

17. MP **Reasoning** Explain how to find a better estimate for 2,589 ÷ 6 than the one shown.

Round 2,589 to 3,000. Estimate 3,000 ÷ 6.

3,000 ÷ 6 = 500, so 2,589 ÷ 6 is about 500.

5.3 Understand Division and Remainders

Use a model to find the quotient and the remainder.

18. $14 \div 4 =$ _____ R _____

19. $28 \div 6 =$ _____ R _____

20. $18 \div 7 =$ _____ R _____

21. $23 \div 3 =$ _____ R _____

22. Modeling Real Life Tours of a factory can have no more than 9 guests. There are 76 guests in line to tour the factory.

- How many tours are full?
- How many tours are needed?
- How many guests are in the last tour?

5.4 Use Partial Quotients

Use partial quotients to divide.

23. $8\overline{)504}$

24. $4\overline{)52}$

25. $7\overline{)119}$

5.5 Use Partial Quotients with a Remainder

Use partial quotients to divide.

26. $5\overline{)82}$

27. $8\overline{)759}$

28. $3\overline{)5{,}468}$

5.6 Divide Two-Digit Numbers by One-Digit Numbers

Divide. Then check your answer.

29. $3\overline{)58}$

30. $4\overline{)90}$

31. $2\overline{)67}$

5.7 Divide Multi-Digit Numbers by One-Digit Numbers

Divide. Then check your answer.

32. $5\overline{)865}$

33. $2\overline{)7{,}532}$

34. $4\overline{)507}$

35. $6\overline{)9{,}127}$

36. $8\overline{)253}$

37. $6\overline{)429}$

5.8 Divide by One-Digit Numbers

Divide. Then check your answer.

38. $3\overline{)91}$

39. $7\overline{)914}$

40. $2\overline{)6{,}075}$

5.9 Problem Solving: Division

41. A young snake sheds its skin every 2 weeks. How many times will the snake shed its skin in 3 years?

6 Factors, Multiples, and Patterns

Chapter Learning Target:
Understand factors, multiples, and patterns.

Chapter Success Criteria:
- I can find the factors of a number.
- I can explain the differences between factors and multiples.
- I can compare the different features of different numbers and shapes.
- I can apply an appropriate strategy to show relationships in numbers and shapes.

- Have you ever watched a basketball game? What was your favorite part?
- An equal number of fans can sit in each bleacher row. How can you use factors to determine the total number of fans that can sit in the bleachers?

Name ______________________________

6 Vocabulary

Review Words
factors
product

Organize It

Complete the graphic organizer.

[ ]

Numbers that are multiplied to get a [ ]

_____ × _____ = 18

_____ × _____ = 24

Define It

Use your vocabulary cards to identify the word. Find the word in the word search.

1. Two factors that, when multiplied, result in a given product

2. A number is _____ by another number when the quotient is a whole number and the remainder is 0.

3. Tells how numbers or shapes in a pattern are related

F	B	I	L	T	D	O	F	R	C
H	A	E	R	I	U	Y	P	E	A
V	Q	C	W	P	E	I	J	L	R
O	U	B	T	S	G	A	D	B	U
R	M	I	L	O	N	B	N	I	M
S	L	A	E	Z	R	L	V	S	T
B	C	F	K	P	D	P	T	I	E
D	E	R	U	L	E	C	A	V	O
O	I	K	V	W	B	U	L	I	F
N	T	P	E	H	C	M	X	D	R

Chapter 6 Vocabulary Cards

composite number	divisible
factor pair	multiple
prime number	rule

A number is divisible by another number when the quotient is a whole number and the remainder is 0.

$48 \div 4 = 12$ R0

So, 48 is divisible by 4.

A whole number greater than 1 with more than two factors

27

The factors of 27 are 1, 3, 9, and 27.

The product of a number and any other counting number.

$1 \times 4 = 4$

$2 \times 4 = 8$

$3 \times 4 = 12$

$4 \times 4 = 16$

multiples of 4

Two factors that, when multiplied, result in a given product

factor pair

$$2 \times 4 = 8$$

factor factor

2 and 4 are a factor pair for 8.

Tells how numbers or shapes in a pattern are related

Rule: Add 3.

3, 6, 9, 12, 15, 18, 21, 24, . . .

Rule: triangle, hexagon, square, rhombus

A number greater than 1 with exactly two factors, 1 and itself

11

The factors of 11 are 1 and 11.

Name ______________________________

Understand Factors 6.1

Learning Target: Use models to find factor pairs.

Success Criteria:

- I can draw area models that show a product.
- I can find the factors of a number.
- I can find the factor pairs for a number.

Explore and Grow

Draw two different rectangles that each have an area of 24 square units. Label their side lengths.

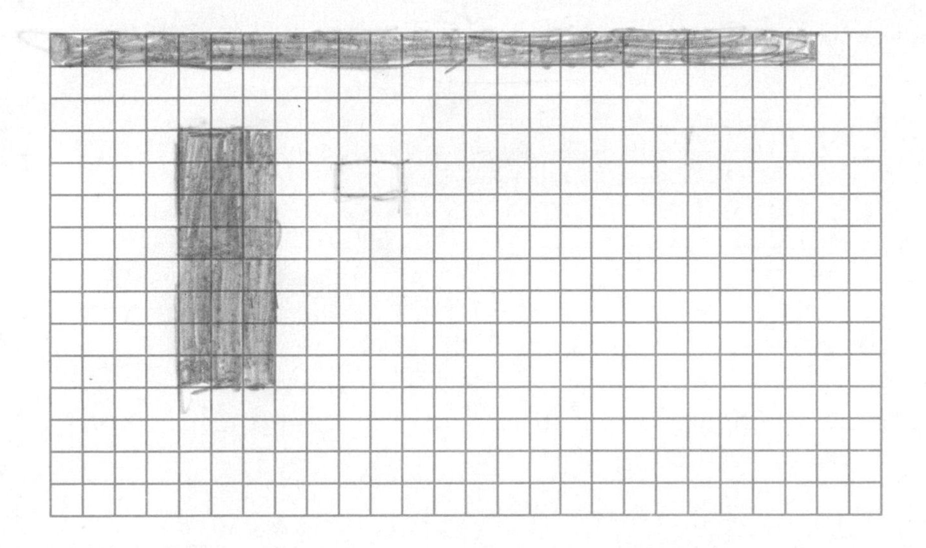

Compare your rectangles to your partner's rectangles. How are they the same? How are they different?

Structure How is each side length related to 24?

Think and Grow: Find Factor Pairs

You can write whole numbers as products of two factors. The two factors are called a **factor pair** for the number.

factor pair

$$2 \times 4 = 8$$

factor factor

2 and 4 are a factor pair for 8.

Example Find the factor pairs for 20.

Find the side lengths of as many different rectangles with an area of 20 square units as possible.

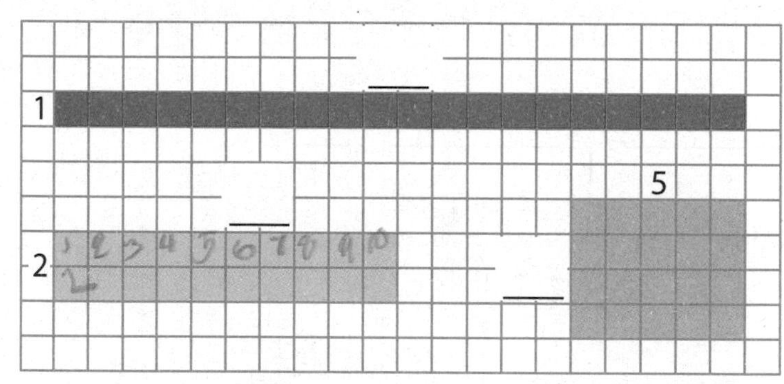

A 4×5 rectangle has the same area as a 5×4 rectangle. Both give the factor pair 4 and 5.

The side lengths of each rectangle are a factor pair.

So, the factor pairs for 20 are ____ and ____, ____ and ____, and ____ and ____.

Show and Grow I can do it!

1. Use the rectangles to find the factor pairs for 12.

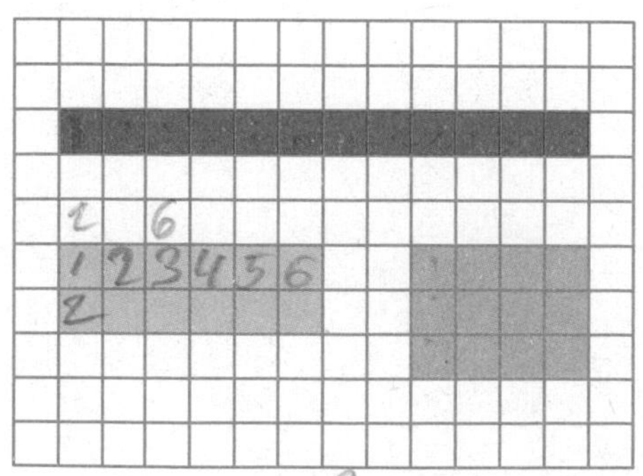

2. Draw rectangles to find the factor pairs for 16.

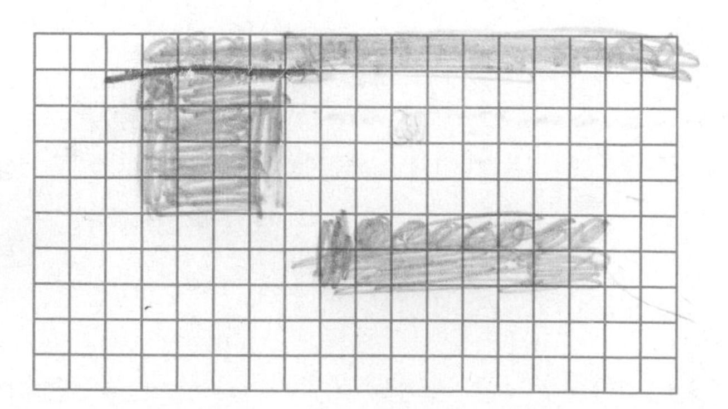

Name ______________________________

Apply and Grow: Practice

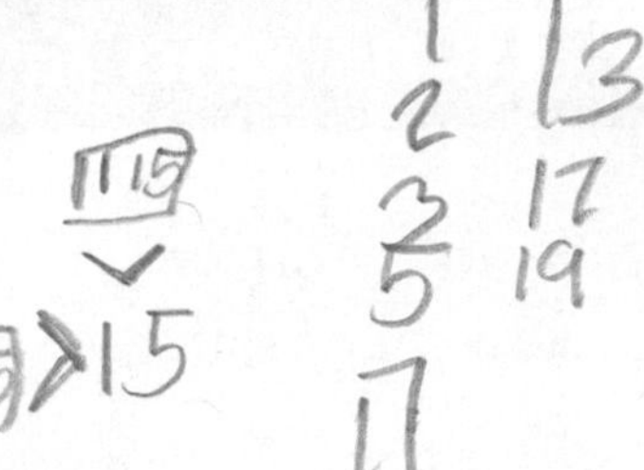

Draw rectangles to find the factor pairs for the number.

3. 14

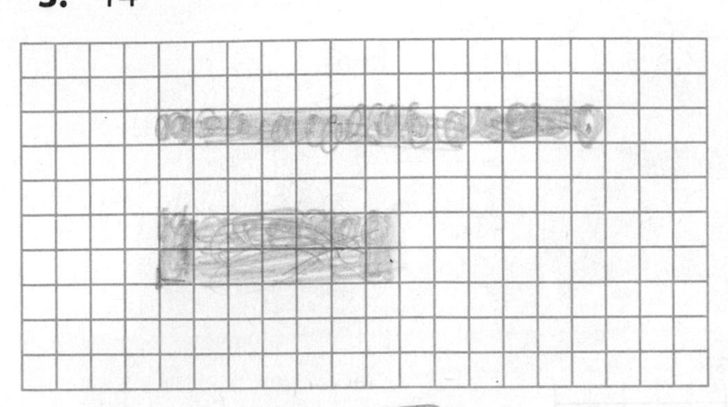

4. 15

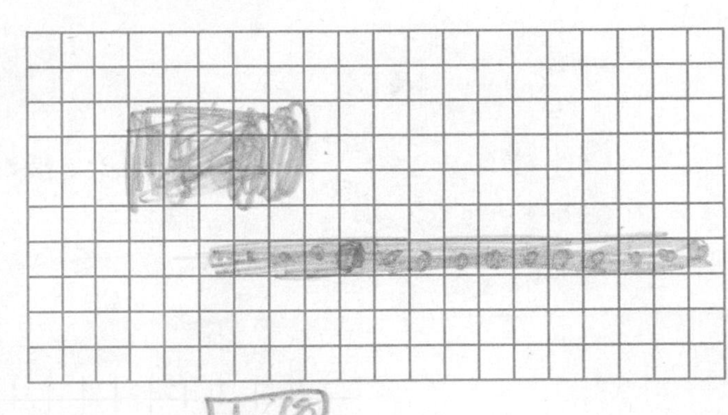

5. 20

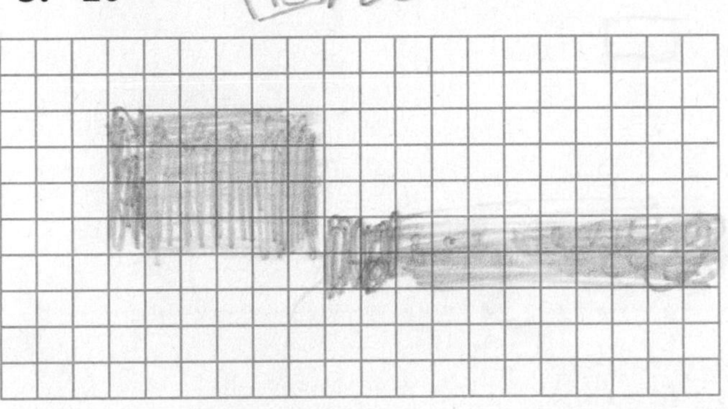

6. 18

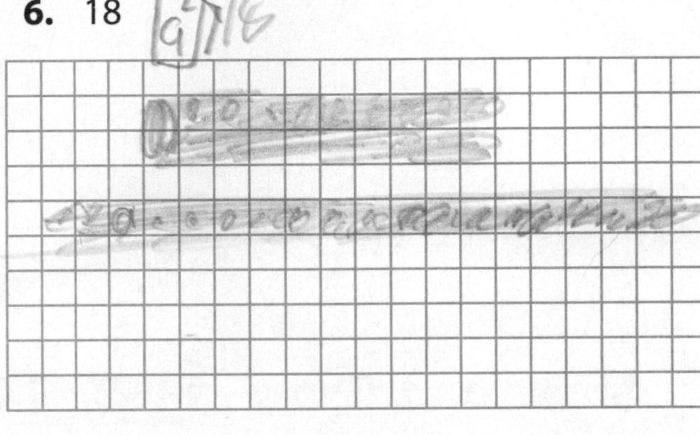

Find the factor pairs for the number.

7. 11

8. 9

9. 4

10. 25

11. 10

12. 40

13. Writing Use the word *factor* to explain one way that 2 and 6 are related.

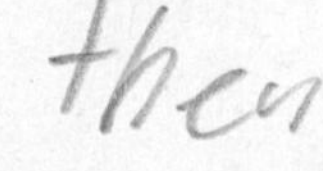

Think and Grow: Modeling Real Life

Example You want to organize 30 pictures into a rectangular array on a wall. How many different arrays can you make?

To find the number of arrays you can make, find the number of factor pairs for 30.

There are ______ factor pairs for 30.

You can use each factor pair to make ______ arrays.

So, there are ______ × ______ = ______ ways to organize the pictures in different arrays.

Show and Grow *I can think deeper!*

14. A city mayor buys 27 solar panels. She wants to organize the panels into a rectangular array. How many different arrays can she make?

15. **DIG DEEPER!** A store owner has 42 masks to hang in a rectangular array on a wall. The owner does not have room for more than 10 masks in each row or column. What are the possible numbers of masks the owner should hang in each row?

16. A teacher wants to set up a chair for each of the 48 students in chorus. He wants to set up the chairs in a rectangular array. He can fit no more than 20 rows and no more than 30 chairs in each row in the room. What are the possible numbers of rows that he could set up?

Name ___________________________

Factors and Divisibility 6.2

Learning Target: Use division to find factor pairs.

Success Criteria:

- I can divide to find factor pairs.
- I can use divisibility rules to find factor pairs.

List any 10 multiples of 3. What do you notice about the sum of the digits in each multiple?

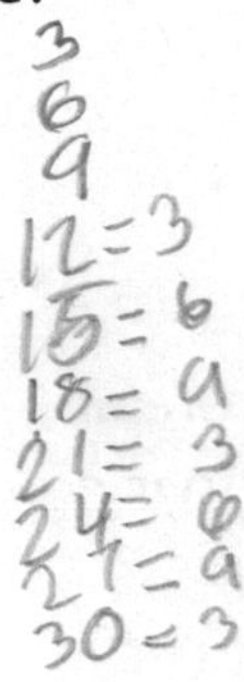

List any 10 multiples of 9. What do you notice about the sum of the digits in each multiple?

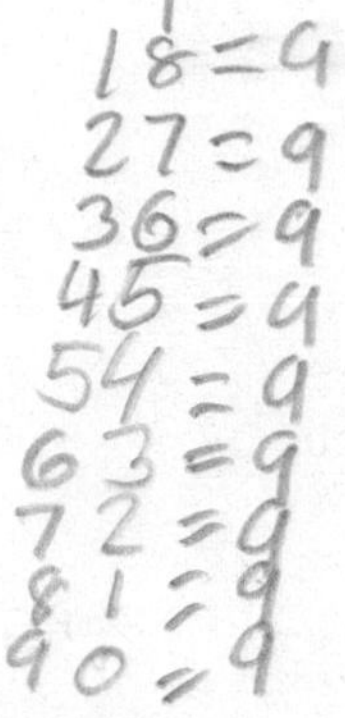

Structure How can you use your observations above to determine whether 3 and 9 are factors of a given number? Explain.

Think and Grow: Find Factors and Factor Pairs

A number is **divisible** by another number when the quotient is a whole number and the remainder is 0.

Some numbers have divisibility rules that you can use to determine whether they are factors of other numbers.

Divisor	Divisibility Rule
2	The number is even.
3	The sum of the digits is divisible by 3.
5	The ones digit is 0 or 5.
6	The number is even and divisible by 3.
9	The sum of the digits is divisible by 9.
10	The ones digit is 0.

Example Find the factor pairs for 48.

A number is divisible by its factors.

Use divisibility rules and division to find the factors of 48.

Divisor	Is the number a factor of 48?	Multiplication Equation
1	Yes, 1 is a factor of every number.	$1 \times$ ____ $= 48$
2	____, 48 is even.	$2 \times$ ____ $= 48$
3	____, $4 + 8 = 12$ is divisible by 3.	$3 \times$ ____ $= 48$
4	____, $48 \div 4 = 12$ R0	$4 \times$ ____ $= 48$
5	____, the ones digit is not 0 or 5.	
6	____, 48 is even and divisible by 3.	$6 \times$ ____ $= 48$
7	____, $48 \div 7 = 6$ R6	
8	____, $48 \div 8 = 6$ R0	$8 \times$ ____ $= 48$

You can stop after checking 8 because the factor pairs start to repeat.

The factors of 48 are ____, ____, ____, ____, ____, ____, ____, ____, ____, and ____.

The factor pairs for 48 are ______________________.

Show and Grow I can do it!

Find the factor pairs for the number.

1. 30

2. 54

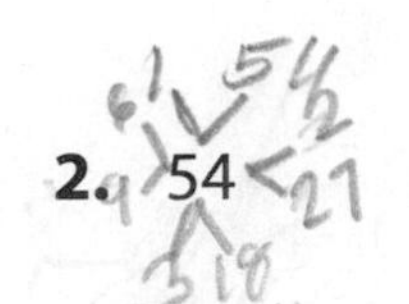

Name ______________________________

Apply and Grow: Practice

Find the factor pairs for the number.

3. 29

4. 50

5. 63

6. 33

7. 60

8. 64

List the factors of the number.

9. 39

10. 44

11. 72

12. 67

13. 42

14. 28

15. **Reasoning** Can an odd number have an even factor? Explain.

16. **Writing** Use the diagram to explain why you do *not* have to check whether any numbers greater than 4 are factors of 12.

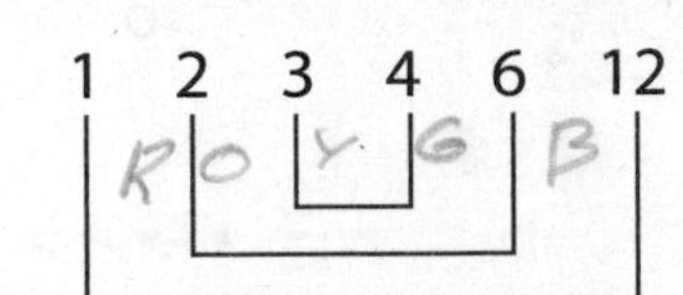

Think and Grow: Modeling Real Life

Example There are 4 classes going on a field trip. The classes will use 3 buses. Can the teachers have an equal number of students on each bus?

Class	Students
1	24
2	23
3	25
4	20

Think: What do you know? What do you need to find? How will you solve?

Step 1: Add to find how many students are going on the field trip.

$24 + 23 + 25 + 20 =$ 92

92 students are going on the field trip.

Step 2: Is the total number of students divisible by the number of buses?

Find the sum of the digits of 92. ____ + ____ = ____

The sum of the digits ________ divisible by 3.

The teachers ________ have an equal number of students on each bus.

Show and Grow *I can think deeper!*

17. A teacher is making a 5-page test with 28 vocabulary problems and 7 reading problems. Can the teacher put an equal number of problems on each page?

18. A relay race is 39 laps long. Each team member must bike the same number of laps. Could a team have 8, 6, or 3 members? Explain.

19. **DIG DEEPER!** You have 63 clay figures to display on 7 shelves. Not all of the shelves need to be used and each shelf can hold no more than 25 figures. Each shelf must have the same number of figures. What are all the ways you could arrange the figures?

63 [79]

Name ______________________________

Relate Factors and Multiples 6.3

Learning Target: Understand the relationship between factors and multiples.

Success Criteria:

- I can tell whether a number is a multiple of another number.
- I can tell whether a number is a factor of another number.
- I can explain the relationship between factors and multiples.

List all factors of 24.

List several multiples of each factor. What number appears in each list?

Number Sense How are factors and multiples related?

Think and Grow: Identify Multiples

A whole number is a multiple of each of its factors.

12 is a multiple of 1, 2, 3, 4, 6, and 12.

$$1 \times 12 = 12$$
$$2 \times 6 = 12$$
$$3 \times 4 = 12$$

Example Is 56 a multiple of 7?

One Way: List multiples of 7.

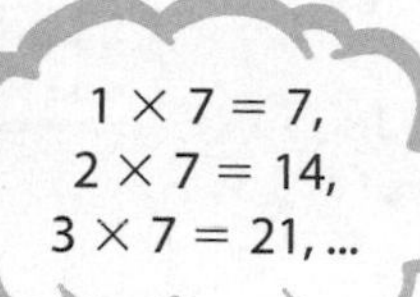

7, 14, 21, _____, _____,

_____, _____, _____

So, 56 _________ a multiple of 7.

Another Way: Use division to determine whether 7 is a factor of 56.

$56 \div 7 =$ _____

7 _______ a factor of 56.

So, 56 _________ a multiple of 7.

Example Is 9 a factor of 64?

One Way: Use divisibility rules to determine whether 9 is a factor of 64.

9 _________ a factor of 64 because

$6 + 4 = 10$ _________ divisible by 9.

Another Way: List multiples of 9.

9, 18, 27, _____, _____, _____, _____, _____

64 _________ a multiple of 9.

So, 9 _______ a factor of 64.

Show and Grow *I can do it!*

1. Is 23 a multiple of 3? Explain.

2. Is 8 a factor of 56? Explain.

Name ________________________________

Apply and Grow: Practice

3. Is 65 a multiple of 5? Explain.

4. Is 14 a multiple of 4? Explain.

5. Is 23 a multiple of 2? Explain.

6. Is 6 a factor of 96? Explain.

7. Is 3 a factor of 82? Explain.

8. Is 9 a factor of 72? Explain.

Tell whether 8 is a multiple or a factor of the number. Write *multiple, factor,* or *both*.

9. 4

10. 8

11. 32

12. **Writing** Use the numbers 6 and 12 to explain how factors and multiples are related.

13. MP **Number Sense** Complete the Venn diagram.

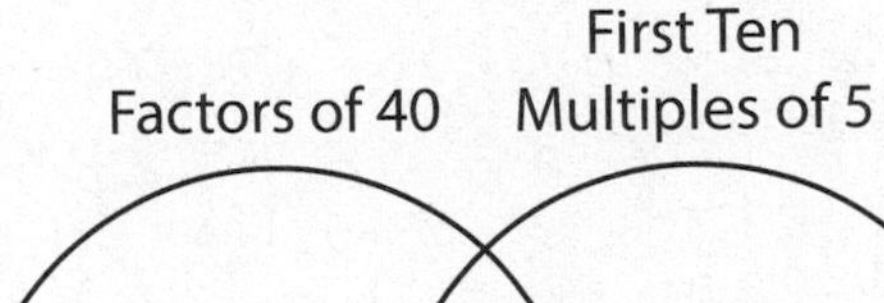

Think and Grow: Modeling Real Life

Example You need 96 balloons for a school dance. Balloons come in packs of 4, packs of 6, and packs of 9. Which packs could you buy so you have no leftover balloons?

Use division to determine whether 96 is a multiple of 4.

$4\overline{)96}$ So, 4 ________ a factor of 96 and 96 ________ a multiple of 4.

Use the divisibility rules to check whether 96 is a multiple of 6.

96 is even and $9 + 6 = 15$ ________ divisible by 3. So, 6 ________ a factor of 96 and 96 ________ a multiple of 6.

Use the divisibility rules to check whether 96 is a multiple of 9.

$9 + 6 = 15$ ________ divisible by 9. So, 9 ________ a factor of 96 and 96 ________ a multiple of 9.

You could buy packs of _____ balloons or packs of _____ balloons.

Show and Grow *I can think deeper!*

14. A teacher needs 88 batteries for science experiments. Batteries are sold in packs of 2, packs of 6, and packs of 8. Which packs could the teacher buy so she has no leftover batteries?

____

15. **DIG DEEPER!** Descartes buys 2 books for a total of \$15. Each book costs a multiple of \$3. How much could each book cost?

16. Newton buys some boxes of dog treats for \$9 each. Descartes buys some bags of cat treats for \$6 each. Newton and Descartes spend the same amount of money on treats. What is the least amount of money they could have spent?

Name ______________________

Homework & Practice 6.3

Learning Target: Understand the relationship between factors and multiples.

Example Is 47 a multiple of 6?

One Way: List multiples of 6.

6, 12, 18, 24, 30, 36, 42, 48

So, 47 is not a multiple of 6.

Another Way: Use division to determine whether 6 is a factor of 47.

$47 \div 6 = 7 \text{ R}5$

6 is not a factor of 47.

So, 47 is not a multiple of 6.

1. Is 16 a multiple of 3? Explain.

2. Is 21 a multiple of 7? Explain.

3. Is 46 a multiple of 2? Explain.

4. Is 5 a factor of 71? Explain.

5. Is 8 a factor of 88? Explain.

6. Is 4 a factor of 80? Explain.

Tell whether 30 is a multiple or a factor of the number. Write *multiple, factor,* or *both*.

7. 30

8. 90

9. 10

Tell whether 10 is a multiple or a factor of the number. Write *multiple, factor,* or *both.*

10. 5

11. 60

12. 10

13. **DIG DEEPER!** Name two numbers that are each a multiple of both 3 and 4. What do you notice about the two multiples?

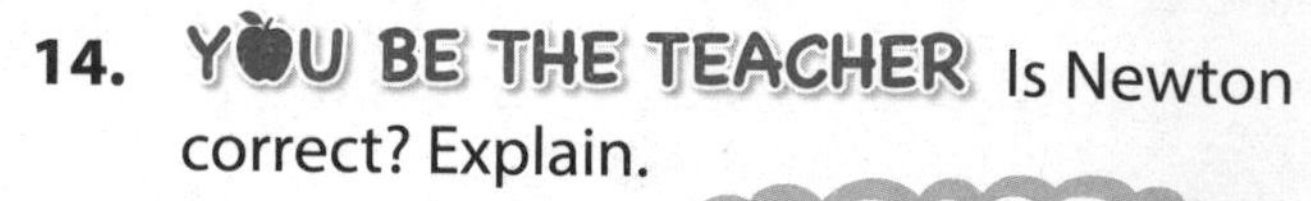

14. **YOU BE THE TEACHER** Is Newton correct? Explain.

15. **MP Logic** A quotient is a multiple of 4. The dividend is a multiple of 8. The divisor is a factor of 6. Write one possible equation for the problem.

16. **Modeling Real Life** Your friend needs to memorize all 50 U.S. state capitals. She wants to memorize the same number of capitals each day. Which numbers of capitals can she memorize each day: 2, 3, 4, or 5?

17. **Modeling Real Life** Zoo keepers plan an enrichment day for the animals every 7 days and bathe the elephants every 2 days. You want to go to the zoo when both events are happening. What other date in May will this happen?

May

Sun	Mon	Tue	Wed	Thu	Fri	Sat
1	2	3	4	5	6	7
8	9	10	11	12	13	14
15	16	17	18	19	20	21
22	23	24	25	26	27	28
29	30	31				

◯ Enrichment Day

▭ Elephant Bath

Review & Refresh

Estimate the sum or difference.

18. $\begin{array}{r} 71{,}606 \\ -\ 49{,}641 \\ \hline \end{array}$

19. $\begin{array}{r} 75{,}294 \\ +\ 36{,}043 \\ \hline \end{array}$

20. $\begin{array}{r} 93{,}294 \\ -\ 40{,}293 \\ \hline \end{array}$

Name ______________________________

Identify Prime and Composite Numbers 6.4

Learning Target: Tell whether a given number is prime or composite.

Success Criteria:

- I can explain what prime and composite numbers are.
- I can identify prime and composite numbers.

Draw as many different rectangles as possible that each have the given area. Label their side lengths.

32 square units	13 square units

Compare the numbers of factors of 32 and 13.

Reasoning Can a whole number have fewer than two factors? exactly two factors? more than two factors?

Think and Grow: Identify Prime and Composite Numbers

A **prime number** is a whole number greater than 1 with exactly two factors, 1 and itself. A **composite number** is a whole number greater than 1 with more than two factors.

Example Tell whether 27 is *prime* or *composite*.

Use divisibility rules.

- 27 is odd, so it ________ divisible by 2 or any other even number.
- 2 + 7 = 9 is divisible by 3,

  so 27 ________ divisible by 3.

27 has factors in addition to 1 and itself.

So, 27 is ____________.

Think: Odd numbers are not divisible by even numbers.

Example Tell whether 11 is *prime* or *composite*.

Use divisibility rules.

- 11 is odd, so it ________ divisible by 2 or any other even number.
- 1 + 1 = 2 is not divisible by 3 or 9,

  so 11 ________ divisible by 3 or 9.
- The ones digit is not 0 or 5,

  so 11 ________ divisible by 5.

11 has exactly two factors, 1 and itself.

So, 11 is ____________.

Show and Grow I can do it!

Tell whether the number is *prime* or *composite*. Explain.

1. 7

2. 12

3. 2

4. 19

5. 45

6. 54

Name ______________________________

Apply and Grow: Practice

Tell whether the number is *prime* or *composite*. Explain.

7. 35

8. 5

9. 23

10. 40

11. 41

12. 81

13. MP **Structure** To create a list of the prime numbers that are less than 100, do the following.

- Place a square around 1. It is neither prime nor composite.
- Circle 2 and cross out all other multiples of 2.
- Circle 3 and cross out all other multiples of 3.
- Circle 5 and cross out all other multiples of 5.
- Circle the next number that is *not* crossed out. This number is prime. Cross out all other multiples of this number.
- Continue until every number is either circled or crossed out.

1	2	3	4	5	6	7	8	9	10
11	12	13	14	15	16	17	18	19	20
21	22	23	24	25	26	27	28	29	30
31	32	33	34	35	36	37	38	39	40
41	42	43	44	45	46	47	48	49	50
51	52	53	54	55	56	57	58	59	60
61	62	63	64	65	66	67	68	69	70
71	72	73	74	75	76	77	78	79	80
81	82	83	84	85	86	87	88	89	90
91	92	93	94	95	96	97	98	99	100

What are the prime numbers that are less than 100? Explain why these numbers were *not* crossed out on the chart.

Think and Grow: Modeling Real Life

Example A museum volunteer has 76 shark teeth to display. Can the volunteer arrange the teeth into a rectangular array with more than 1 row and more than 1 tooth in each row? Explain.

Use divisibility rules to determine whether 76 is prime or composite.

76 is even, so it __________ divisible by 2.

76 has factors in addition to 1 and itself.

So, 76 is ____________.

The volunteer __________ arrange the teeth into a rectangular array with more than 1 row and more than 1 tooth in each row.

Explain:

Show and Grow *I can think deeper!*

14. A teacher has 29 students in class. Can the teacher separate the students into equal groups? Explain.

15. A band instructor wants to have several ways to organize band members into rectangular arrays on the field for a performance. Should the instructor have 89 members or 99 members on the field? Explain.

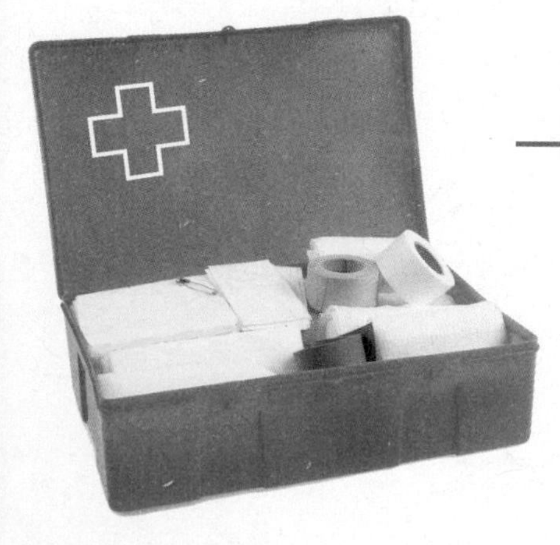

16. **DIG DEEPER!** A paramedic is arranging bandages into 4 bins. An equal number of bandages are in each bin. Did the paramedic arrange a prime number or a composite number of bandages? Explain.

Name ______________________________

Learning Target: Tell whether a given number is prime or composite.

Example Tell whether 43 is *prime* or *composite*.

Use divisibility rules.

- 43 is odd, so it <u>is not</u> divisible by 2 or any other even number.
- $4 + 3 = 7$ is not divisible by 3 or 9, so 43 <u>is not</u> divisible by 3 or 9.
- The ones digit is not 0 or 5, so 43 <u>is not</u> divisible by 5.

43 has exactly two factors, 1 and itself. So, 43 is <u>prime</u>.

Example Tell whether 55 is *prime* or *composite*.

Use divisibility rules.

- 55 is odd, so it <u>is not</u> divisible by 2 or any other even number.
- $5 + 5 = 10$ is not divisible by 3 or 9, so 55 <u>is not</u> divisible by 3 or 9.
- The ones digit is 0 or 5, so 55 <u>is</u> divisible by 5.

55 has factors in addition to 1 and itself. So, 55 is <u>composite</u>.

Tell whether the number is *prime* or *composite*. Explain.

1. 3

2. 27

3. 46

4. 17

5. 53

6. 63

7. 29

8. 31

9. 75

10. **DIG DEEPER!** Can a number be both prime and composite? Explain.

11. **MP Logic** Your friend is thinking of a prime number between 60 and 80. The tens digit is one less than the ones digit. What is the number?

MP Number Sense Write *true* or *false* for the statement. If false, provide an example to support your answer.

12. All odd numbers are prime. ____________

____

13. All even numbers, except 2, are composite. ____________

____

14. A composite number cannot have exactly three factors. ____________

15. **Modeling Real Life** There are 43 students trying out for a basketball team. Can the coach separate the students into equal groups? Explain.

16. **Modeling Real Life** Which planet does *not* have a prime number of rings?

Number of Rings Around a Planet	
Jupiter	◯ ◖
Saturn	◯ ◯ ◯ ◯ ◖
Uranus	◯ ◯ ◯ ◯ ◯ ◯ ◖
Neptune	◯ ◯ ◖

Each ◯ = 2 rings.

Review & Refresh

Use properties to find the product. Explain your reasoning.

17. $4 \times 9 \times 25$

18. 405×3

19. 698×7

Name ________________________________

Number Patterns 6.5

Learning Target: Create and describe number patterns.

Success Criteria:

- I can create a number pattern given a number rule.
- I can describe features of a number pattern.

Shade every third square in the table.

1	2	3	4	5	6	7	8	9	10
11	12	13	14	15	16	17	18	19	20
21	22	23	24	25	26	27	28	29	30
31	32	33	34	35	36	37	38	39	40
41	42	43	44	45	46	47	48	49	50
51	52	53	54	55	56	57	58	59	60

Write the shaded numbers. What patterns do you see?

What other patterns do you see in the table?

Structure Circle every fourth square in the table. Write the circled numbers. What patterns do you see?

Think and Grow: Create Number Patterns

A **rule** tells how numbers or shapes in a pattern are related.

Example Use the rule "Add 3." to create a number pattern. The first number in the pattern is 3. Then describe another feature of the pattern.

Create the pattern.

+ 3 + 3 + 3 + 3 + 3

3, 6, ____, ____, ____, ____, . . .

The numbers in the pattern are multiples of ____.

Example Use the rule "Multiply by 2." to create a number pattern. The first number in the pattern is 10. Then describe another feature of the pattern.

Create the pattern.

× 2 × 2 × 2 × 2 × 2

10, 20, ____, ____, ____, ____, . . .

The ones digit of each number in the pattern is ____.

When describing another feature of the pattern, look at the ones digits or the tens digits. Are all of the numbers even or odd?

Show and Grow I can do it!

Write the first six numbers in the pattern. Then describe another feature of the pattern.

1. Rule: Add 5.
First number: 1

1, ____, ____, ____, ____, ____

2. Rule: Multiply by 3.
First number: 3

3, ____, ____, ____, ____, ____

3. Rule: Subtract 2.
First number: 20

4. Rule: Divide by 2.
First number: 256

Name ________________________________

Apply and Grow: Practice

Write the first six numbers in the pattern. Then describe another feature of the pattern.

5. Rule: Add 11.
First number: 11

6. Rule: Multiply by 4.
First number: 4

7. Rule: Subtract 3.
First number: 21

8. Rule: Divide by 3.
First number: 729

9. Rule: Add 9.
First number: 8

10. Rule: Multiply by 5.
First number: 5

Open-Ended Use the rule to generate a pattern of four numbers.

11. Rule: Multiply by 2.

12. Rule: Subtract 9.

13. Rule: Divide by 4.

14. Rule: Add 7.

15. MP **Patterns** Write a rule for the pattern below. Then write a different pattern that follows the same rule.

3, 6, 12, 24, 48

16. MP **Reasoning** What is the missing number in the pattern? Explain.

39, 37, 35, ______, 31, 29

Think and Grow: Modeling Real Life

Example A presidential election is held every 4 years. There was a presidential election in 2016. How many presidential elections will occur between 2017 and 2030?

The rule is to add 4 years to each presidential election year. Start with 2016. Then count the years in the pattern that are between 2017 and 2030.

+4 +4 +4 +4

2016, ____, ____, ____, ____

____ presidential elections will occur between 2017 and 2030.

Show and Grow *I can think deeper!*

17. The pattern of animals on a Chinese calendar repeats every 12 years. The year 2000 was the year of the dragon. How many times will the year of the dragon occur between 2001 and 2100?

18. A robotics team raised $25 the first month of school. Each month of school, the team wants to raise 2 times as much money as the month before. How much money should they raise in the fifth month of school?

19. **DIG DEEPER!** You start with 128 pictures on your tablet. You take 6 pictures and delete 3 pictures each day. How many pictures do you have on your tablet after 6 days?

Name ________________________________

Homework & Practice 6.5

Learning Target: Create and describe number patterns.

Example Use the rule "Subtract 6." to create a number pattern. The first number in the pattern is 100. Then describe another feature of the pattern.

When describing another feature of the pattern, look at the ones digits or the tens digits. Are all of the numbers even or odd?

Create the pattern.

− 6 − 6 − 6 − 6 − 6

100, 94, <u>88</u>, <u>82</u>, <u>76</u>, <u>70</u>, . . .

The numbers in the pattern are <u>even</u>.

Example Use the rule "Divide by 5." to create a number pattern. The first number in the pattern is 3,125. Then describe another feature of the pattern.

Create the pattern.

÷ 5 ÷ 5 ÷ 5 ÷ 5 ÷ 5

3,125, 625, <u>125</u>, <u>25</u>, <u>5</u>, <u>1</u>, . . .

The ones digit of each number in the pattern except the last number is <u>5</u>.

Write the first six numbers in the pattern. Then describe another feature of the pattern.

1. Rule: Subtract 8.
First number: 88

88, _____, _____, _____, _____, _____

2. Rule: Multiply by 10.
First number: 2

2, _____, _____, _____, _____, _____

3. Rule: Add 9.
First number: 17

4. Rule: Divide by 2.
First number: 1,600

Open-Ended Use the rule to generate a pattern of four numbers.

5. Rule: Divide by 5.

6. Rule: Add 8.

7. Rule: Multiply by 9.

8. Rule: Subtract 3.

9. **MP Structure** List the first ten multiples of 9. What patterns do you notice with the digits in the ones place? in the tens place?

Does this pattern continue beyond the tenth number in the pattern?

10. **Modeling Real Life** It takes the moon about 28 days to orbit Earth. How many times will the moon orbit Earth in 1 year?

11. **DIG DEEPER!** In each level of a video game, you can earn up to 10 points and lose up to 3 points. Your friend earns 9 points in the first level. If he earns and loses the maximum number of points each level, how many total points will he have after level 6?

Review & Refresh

Find the product.

12. $14 \times 23 =$ ______

13. $48 \times 60 =$ ______

14. $55 \times 31 =$ ______

Name ______________________________

Shape Patterns 6.6

Learning Target: Create and describe shape patterns.

Success Criteria:

- I can create a shape pattern given a rule.
- I can find the shape at a given position in a pattern.
- I can describe features of a shape pattern.

Create a rule using 3 different shapes. Draw the first six shapes in the pattern.

What is the next shape in the pattern?

What is the 9th shape in the pattern? Explain.

What is the 99th shape? 1,000th shape? Explain.

Structure You want to show the first 40 shapes in the pattern above. Without modeling, how many of each shape do you think you will need?

Think and Grow: Create Shape Patterns

Example Create a shape pattern by repeating the rule "triangle, hexagon, square, rhombus." What is the 42nd shape in the pattern?

Create the pattern.

42 ÷ 4 is 10 R2, so when the pattern repeats 10 times,

the 40th shape is a ___________. So, the 41st shape is a

___________ and the 42nd shape is a ___________.

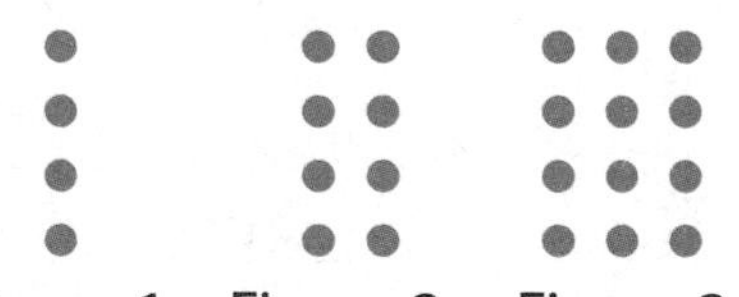

Example Describe the dot pattern. How many dots are in the 25th figure?

Figure 1 Figure 2 Figure 3

Figure 1 has 1 column of 4 dots, so it has 1 × _____ = _____ dots.

Figure 2 has 2 columns of 4 dots, so it has 2 × _____ = _____ dots.

Figure 3 has 3 columns of 4 dots, so it has 3 × _____ = _____ dots.

The 25th figure has _____ columns of 4 dots, so it has _____ × _____ = _____ dots.

Show and Grow *I can do it!*

1. Extend the pattern of shapes by repeating the rule "square, trapezoid, triangle, hexagon, triangle." What is the 108th shape in the pattern?

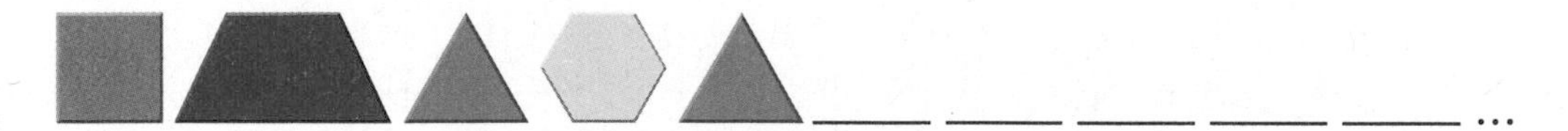

2. Describe the dot pattern. How many dots are in the 76th figure?

Figure 1 Figure 2 Figure 3

Name ____________________

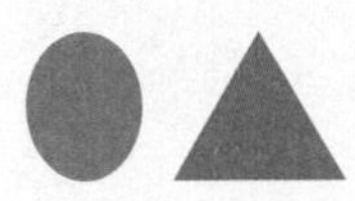

Apply and Grow: Practice

3. Extend the pattern of shapes by repeating the rule "oval, triangle." What is the 55th shape?

4. Extend the pattern of symbols by repeating the rule "add, subtract, multiply, divide." What is the 103rd symbol?

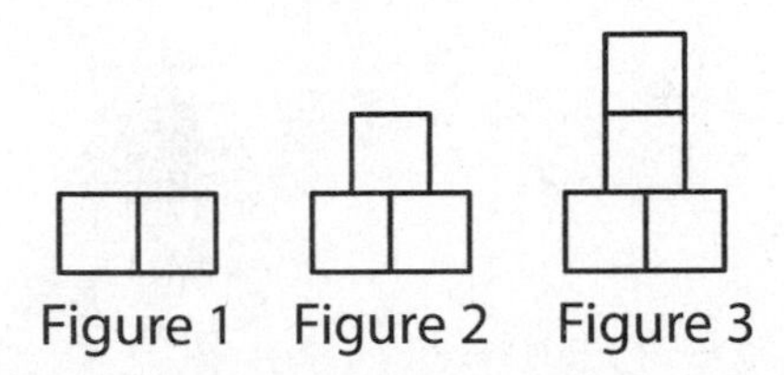

5. Describe the pattern. How many squares are in the 24th figure?

Figure 1 Figure 2 Figure 3

6. Describe the pattern of the small triangles. How many small triangles are in the 10th figure?

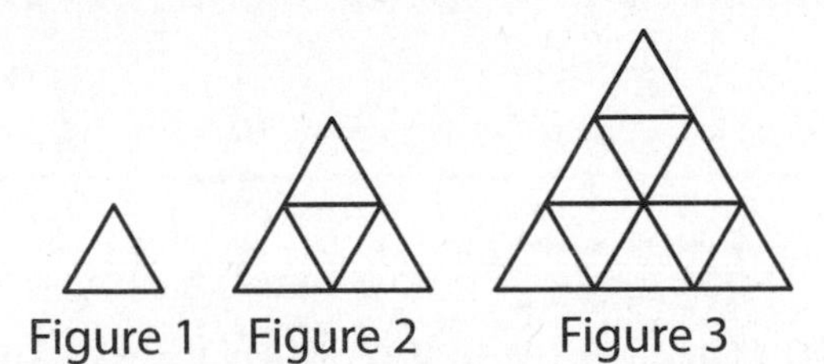

7. MP **Structure** Make a shape pattern that uses twice as many squares as triangles.

8. MP **Number Sense** Which shape patterns have a heart as the 12th shape?

Think and Grow: Modeling Real Life

Example You make a necklace with cube, hexagon, and star beads. You string the beads in a pattern. You use the rule "cube, star, cube, hexagon." It takes 64 beads to complete the necklace. How many times do you repeat the pattern?

Divide the number of beads it takes to complete the necklace by the number of beads in the rule. There are 4 beads in the rule.

$4\overline{)64}$

You repeat the pattern _____ times.

Show and Grow *I can think deeper!*

9. The path on a board game uses the rule "red, green, pink, yellow, blue." There are 55 spaces on the game board. How many times does the pattern repeat?

10. You make a walkway in a garden using different-shaped stepping stones. You use the rule "square, circle, square, hexagon." You use 24 square stepping stones. How many circle and hexagon stepping stones do you use altogether? How many stones do you use in all?

11. **DIG DEEPER!** You make a rectangular picture frame using square tiles. The picture frame is 12 tiles long and 8 tiles wide. You arrange the tiles in a pattern. You use the rule "red, orange, yellow." How many of each color tile do you use?

Name ______________________________

Homework & Practice 6.6

Learning Target: Create and describe shape patterns.

Example Create a shape pattern by repeating the rule "hexagon, octagon." What is the 29th shape in the pattern?

Create the pattern.

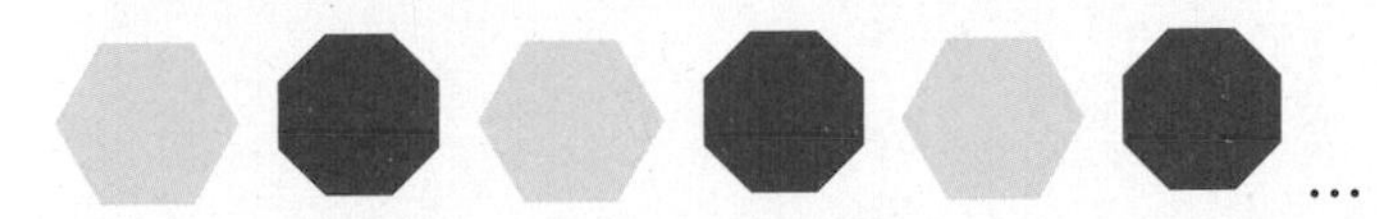

29 divided by 2 is 14 R1, so when the pattern repeats 14 times the 28th shape is an __octagon__. So, the 29th shape is a __hexagon__.

Example Describe the dot pattern. How many dots are in the 64th figure?

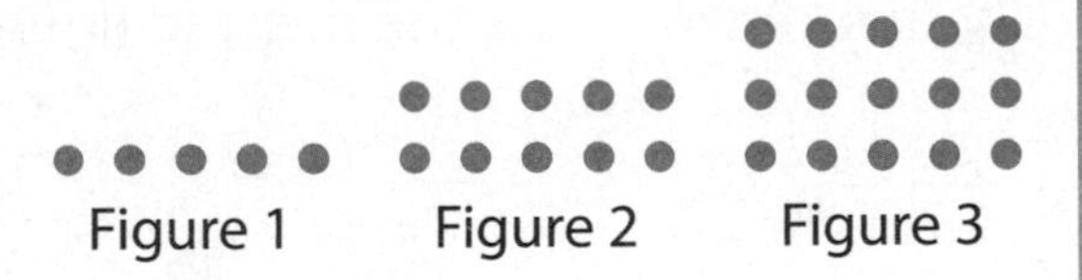

Figure 1 has 1 row of 5 dots, so it has 1 × __5__ = __5__ dots.

Figure 2 has 2 rows of 5 dots, so it has 2 × __5__ = __10__ dots.

Figure 3 has 3 rows of 5 dots, so it has 3 × __5__ = __15__ dots.

The 64th figure has __64__ rows of 5 dots, so it has __64__ × __5__ = __320__ dots.

1. Extend the pattern of shapes by repeating the rule "up, right, down, left." What is the 48th shape in the pattern?

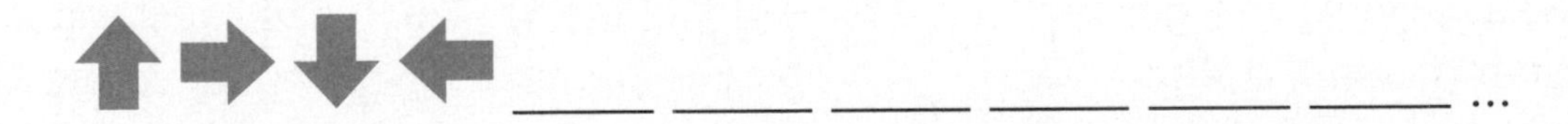

2. Extend the pattern of shapes by repeating the rule "small circle, medium circle, large circle." What is the 86th shape in the pattern?

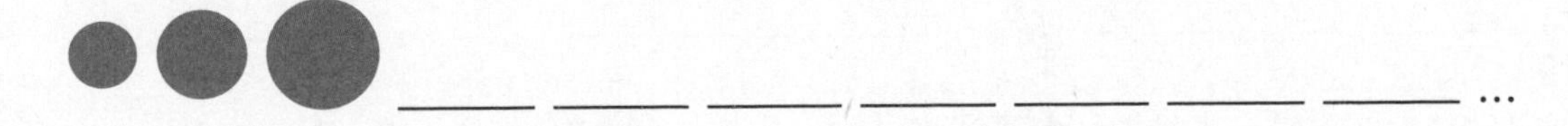

3. Describe the dot pattern. How many dots are in the 113th figure?

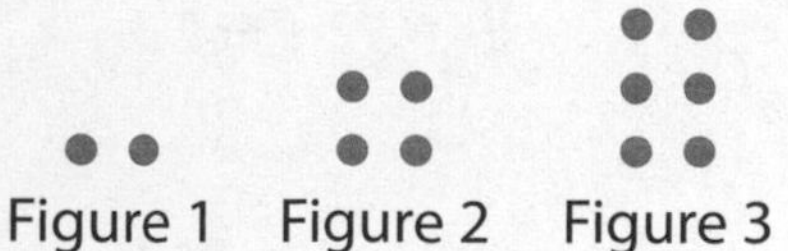

4. **YOU BE THE TEACHER** You and your friend each create a shape pattern with 100 shapes. Your friend says both patterns will have the same number of circles. Is your friend correct? Explain.

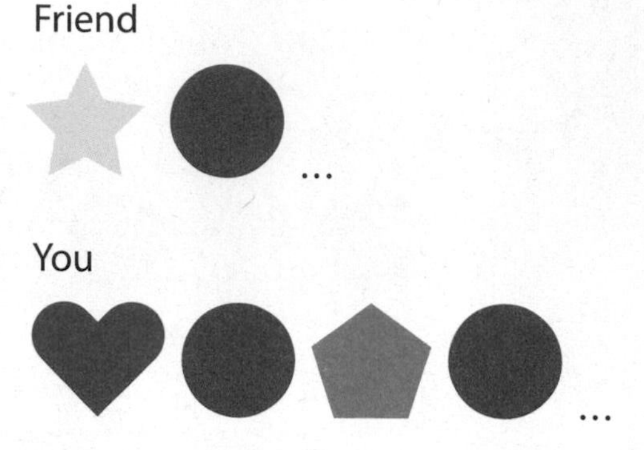

5. **Structure** Draw the missing figure in the pattern. Explain the pattern.

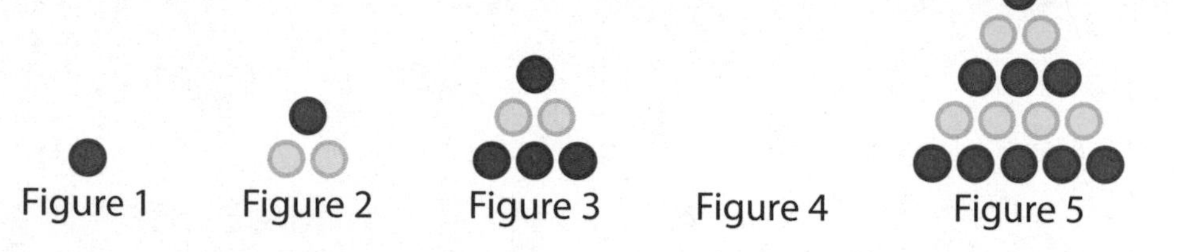

6. **Reasoning** Newton uses the rule "bone, bone, paw print" to make a shape pattern. He wants the pattern to repeat 8 times. How many bones will be in Newton's pattern?

7. **Modeling Real Life** The black keys on a piano follow the pattern "two black keys, three black keys." There are 36 black keys on a standard piano. How many times does this entire pattern repeat?

Review & Refresh

Find the quotient.

8. $30 \div 5 =$ ______	**9.** $360 \div 9 =$ ______	**10.** $6,400 \div 8 =$ ______
11. $140 \div 2 =$ ______	**12.** $4,200 \div 7 =$ ______	**13.** $40 \div 2 =$ ______

Name ______________________________

Performance Task 6

You play basketball in a youth basketball program.

1. There are 72 players in the program. Each team needs an equal number of players and must have at least 5 players. What are two different ways the teams can be made?

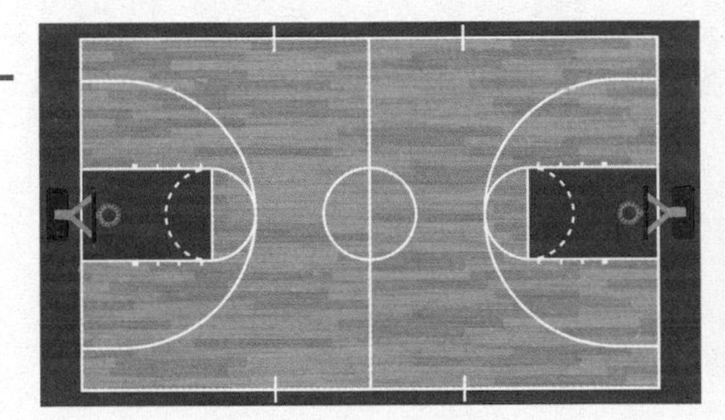

2. The width of a basketball court is 42 feet and the length is 74 feet. You run around the perimeter of the court 4 times to warm up. How many feet do you run?

3. You and your friend are on the same team. You played your first game last week.

   a. Your team scored 4 more points than the other team. The total number of points scored by both teams was 58. How many points did your team score?

   b. You and your friend scored the same number of points. You made 2-point shots and your friend made 3-point shots. What could be the greatest number of points you and your friend each scored?

4. Your team uses the pattern below to decide which jersey color to wear to each game. Which color jersey will your team wear on the 20th game?

Game 1

Game 2

Game 3

Game 4

Multiple Lineup

Directions:

1. Players take turns rolling a die.
2. On your turn, place a counter on a multiple of the number of your roll. If there is not a multiple of the number of your roll, you lose your turn.
3. The first player to create a line of 5 in a row, horizontally, vertically, or diagonally, wins!

30	18	9	16	36
15	4	10	44	17
25	42	7	80	21
6	75	22	45	56
11	27	12	95	24

Name ______________________________

Chapter Practice 6

6.1 Understand Factors

1. Use the rectangles to find the factor pairs for 6.

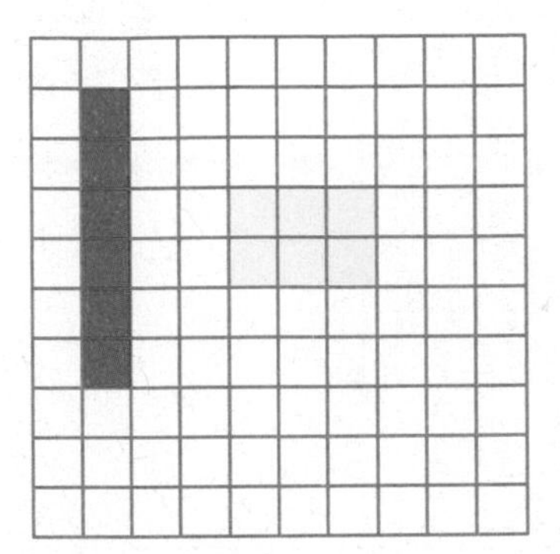

2. Draw rectangles to find the factor pairs for 12.

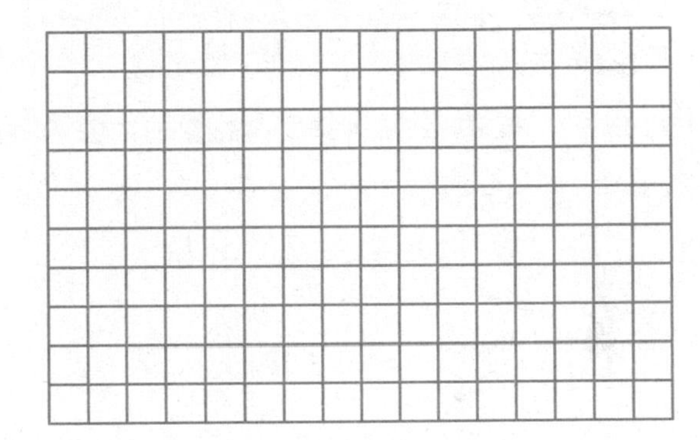

Find the factor pairs for the number.

3. 17

4. 10

5. 21

6. 20

7. 36

8. 50

6.2 Factors and Divisibility

Find the factor pairs for the number.

9. 16

10. 24

11. 56

List the factors of the number.

12. 25

13. 60

14. 72

15. MP **Number Sense** Which numbers have 3 as a factor?

56 21 36 48 93 71

6.3 Relate Factors and Multiples

16. Is 54 a multiple of 3? Explain.

17. Is 45 a multiple of 7? Explain.

18. Is 2 a factor of 97? Explain.

19. Is 5 a factor of 60? Explain.

Tell whether 20 is a multiple or a factor of the number. Write *multiple, factor,* or *both*.

20. 60

21. 4

22. 20

23. MP **Number Sense** Name two numbers that are each a multiple of both 5 and 2. What do you notice about the two multiples?

24. MP **Logic** A quotient is a multiple of 5. The dividend is a multiple of 4. The divisor is a factor of 8. Write one possible equation for the problem.

6.4 Identify Prime and Composite Numbers

Tell whether the number is *prime* or *composite*. Explain.

25. 5

26. 25

27. 51

28. 21

29. 50

30. 83

31. **Modeling Real Life** A prime number of students have which type of fingerprint?

Loop

Arch

Whorl

Type of Fingerprint	
Loop	◯ ◯ ◯ ◯ ◐
Arch	◯ ◯ ◯ ◯ ◯ ◯
Whorl	◯ ◯ ◯ ◐

Each ◯ = 2 students.

6.5 Number Patterns

Write the first six numbers in the pattern. Then describe another feature of the pattern.

32. Rule: Subtract 11.
First number: 99

99, _____, _____, _____, _____, _____

33. Rule: Multiply by 5.
First number: 10

10, _____, _____, _____, _____, _____

34. Rule: Add 8.
First number: 15

35. Rule: Divide by 4.
First number: 4,096

Open-Ended Use the rule to generate a pattern of four numbers.

36. Rule: Divide by 2.

37. Rule: Add 3.

38. Rule: Multiply by 10.

39. Rule: Subtract 6.

6.6 Shape Patterns

40. Extend the pattern of shapes by repeating the rule "trapezoid, circle." What is the 57th shape in the pattern?

 ______ ______ ______ ______ ______ ______ ______ ______ . . .

41. Extend the pattern of shapes by repeating the rule "top left, top right, bottom right, bottom left." What is the 102nd shape in the pattern?

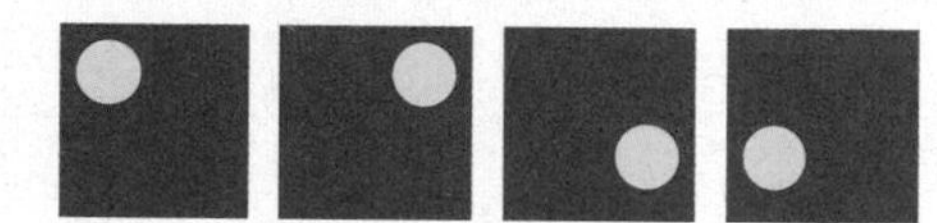 ______ ______ ______ ______ ______ ______ . . .

42. Describe the pattern. How many squares are in the 61st figure?

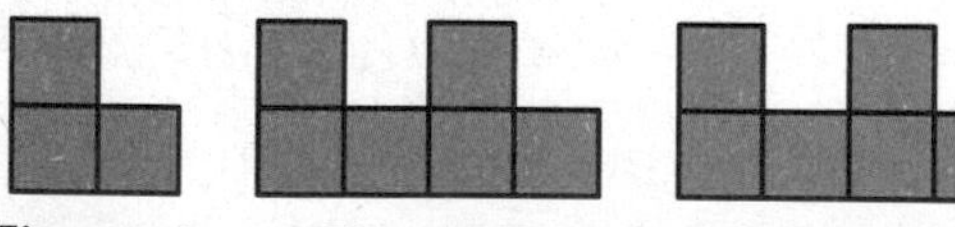

Figure 1 Figure 2

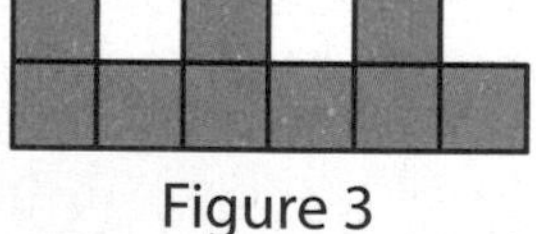

Figure 3

43. MP **Structure** Draw the missing figure in the pattern. Explain the pattern.

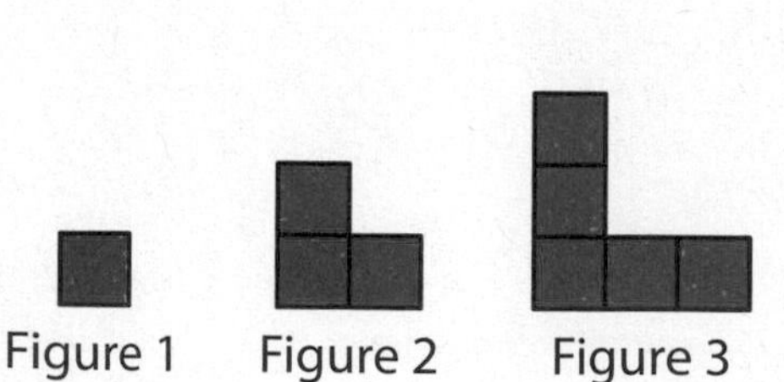

Figure 1 Figure 2 Figure 3

Figure 4

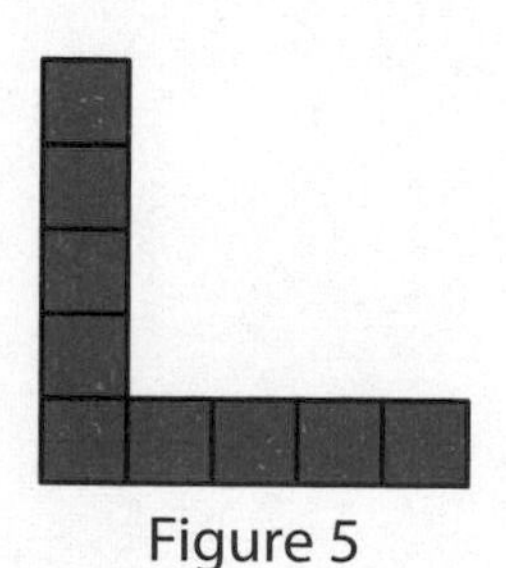

Figure 5

7 Understand Fraction Equivalence and Comparison

Chapter Learning Target:
Understand fractions.

Chapter Success Criteria:
- I can define equivalent fractions.
- I can explain how multiplication can be used to find equivalent fractions.
- I can compare the numerators and denominators of two fractions.
- I can find the factors of a number.

- What are your favorite colors?
- How can you use fractions to compare the amounts of each color of paint you use?

Name ________________________________

7 Vocabulary

Review Words

denominator
fraction
numerator

Organize It

Use the review words to complete the graphic organizer.

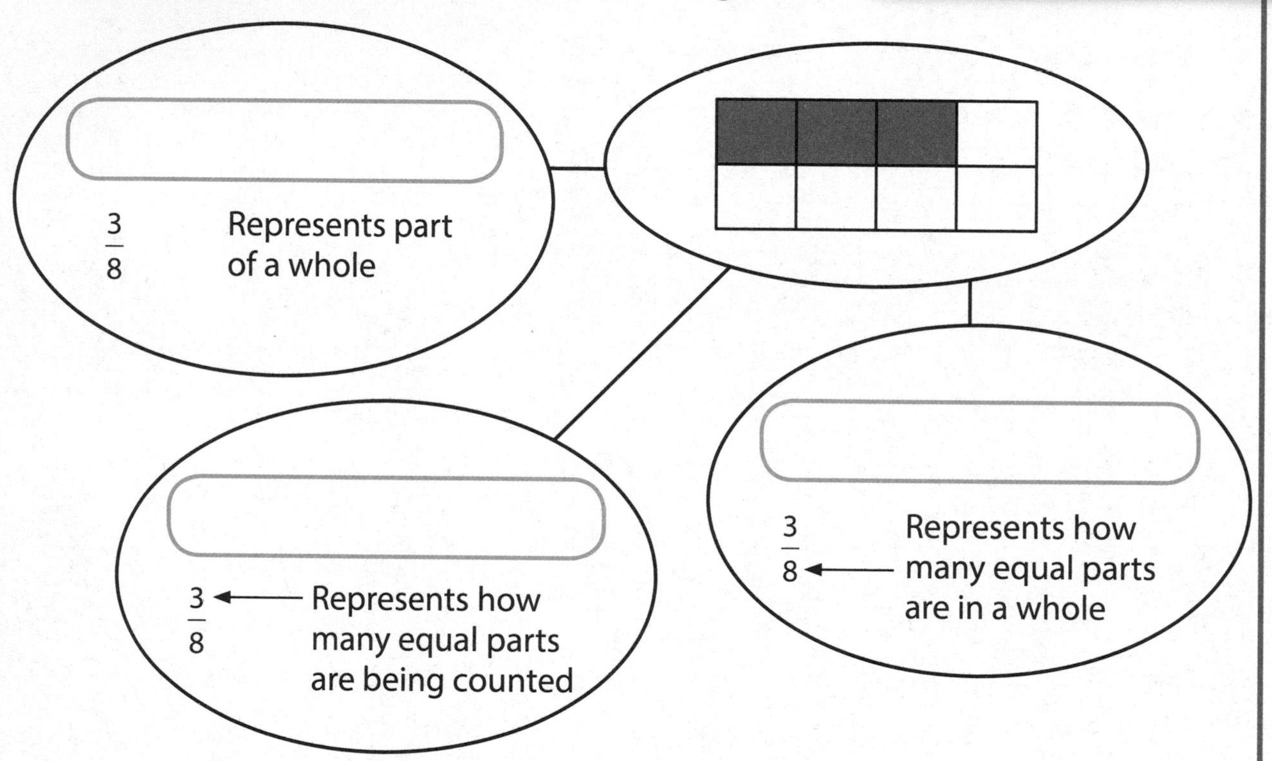

Define It

Use your vocabulary cards to complete each definition.

1. benchmark: A commonly used ________________ that you can use to ________________ other numbers

2. common factor: A ________________ that is ________________ by two or more given ________________

3. equivalent fractions: Two or more ________________ that name the ________________ part of a ________________

Chapter 7 Vocabulary Cards

benchmark

common factor

equivalent

equivalent fractions

A factor that is shared by two or more given numbers

Factors of 8: (1), (2), (4), 8

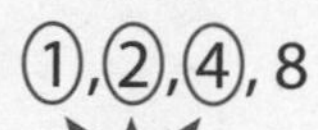

common factors

Factors of 12: (1), (2), 3, (4), 6, 12

A commonly used number that you can use to compare other numbers

Examples: $\frac{1}{2}$, 1

Two or more fractions that name the same part of a whole

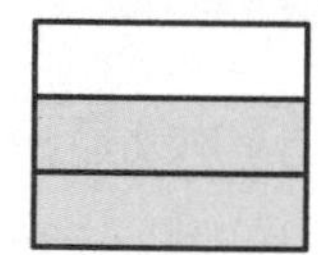

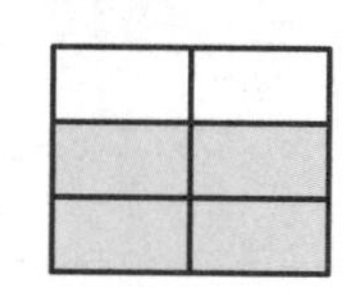

$\frac{2}{3} = \frac{4}{6}$

Having the same value

$\frac{8}{8} = 1$

$3 = \frac{3}{1}$

$2 = \frac{4}{2} = \frac{6}{3}$

Name ______________________________

Model Equivalent Fractions 7.1

Learning Target: Model and write equivalent fractions.

Success Criteria:
- I can use an area model to find equivalent fractions.
- I can use a number line to find equivalent fractions.
- I can write equivalent fractions.

Explore and Grow

Use the model to write fractions that are the same size as $\frac{1}{2}$.

1 whole											
$\frac{1}{2}$	$\frac{1}{2}$										
$\frac{1}{3}$	$\frac{1}{3}$	$\frac{1}{3}$									
$\frac{1}{4}$	$\frac{1}{4}$	$\frac{1}{4}$	$\frac{1}{4}$								
$\frac{1}{5}$	$\frac{1}{5}$	$\frac{1}{5}$	$\frac{1}{5}$	$\frac{1}{5}$							
$\frac{1}{6}$	$\frac{1}{6}$	$\frac{1}{6}$	$\frac{1}{6}$	$\frac{1}{6}$	$\frac{1}{6}$						
$\frac{1}{8}$	$\frac{1}{8}$	$\frac{1}{8}$	$\frac{1}{8}$	$\frac{1}{8}$	$\frac{1}{8}$	$\frac{1}{8}$	$\frac{1}{8}$				
$\frac{1}{10}$	$\frac{1}{10}$	$\frac{1}{10}$	$\frac{1}{10}$	$\frac{1}{10}$	$\frac{1}{10}$	$\frac{1}{10}$	$\frac{1}{10}$	$\frac{1}{10}$	$\frac{1}{10}$		
$\frac{1}{12}$	$\frac{1}{12}$	$\frac{1}{12}$	$\frac{1}{12}$	$\frac{1}{12}$	$\frac{1}{12}$	$\frac{1}{12}$	$\frac{1}{12}$	$\frac{1}{12}$	$\frac{1}{12}$	$\frac{1}{12}$	$\frac{1}{12}$

Reasoning Can you write a fraction with a denominator of 12 that is the same size as $\frac{2}{3}$? Explain.

Think and Grow: Model Equivalent Fractions

Two or more numbers that have the same value are **equivalent**. Two or more fractions that name the same part of a whole are **equivalent fractions**. Equivalent fractions name the same point on a number line.

Example Use models to find equivalent fractions for $\frac{2}{3}$.

One Way: Draw models that show the same whole divided into different numbers of parts.

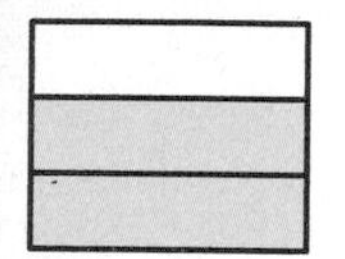
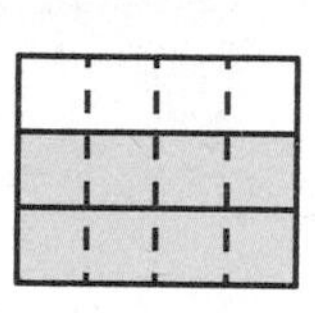
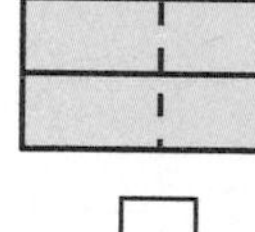

$\frac{2}{3}$ $\frac{\square}{\square}$ $\frac{\square}{\square}$

$\frac{2}{3}$ is equivalent to $\frac{\square}{\square}$ and $\frac{\square}{\square}$.

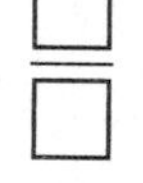
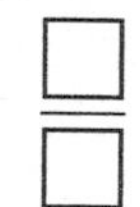

So, $\frac{2}{3} = \frac{\square}{\square}$ and $\frac{2}{3} = \frac{\square}{\square}$.

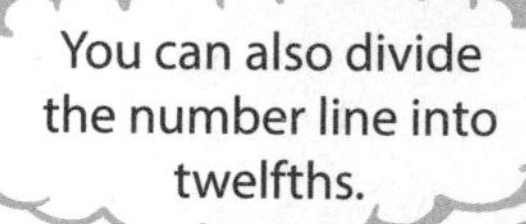

Another Way: Use a number line.

Step 1: Plot $\frac{2}{3}$ on a number line.

Step 2: Divide the number line into sixths and label the tick marks.

$\frac{1}{6}$	$\frac{1}{6}$	$\frac{1}{6}$	$\frac{1}{6}$	$\frac{1}{6}$	$\frac{1}{6}$
$\frac{1}{3}$		$\frac{1}{3}$		$\frac{1}{3}$	

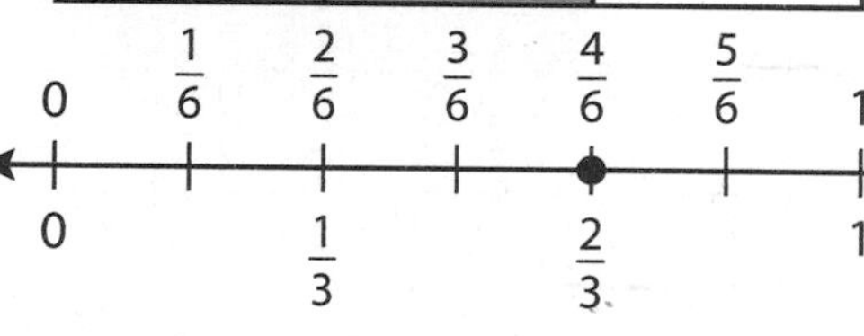

The fractions that name the same point are $\frac{2}{3}$ and $\frac{\square}{\square}$. So, $\frac{2}{3} = \frac{\square}{\square}$.

Show and Grow *I can do it!*

1. Use the model to find an equivalent fraction for $\frac{2}{5}$.

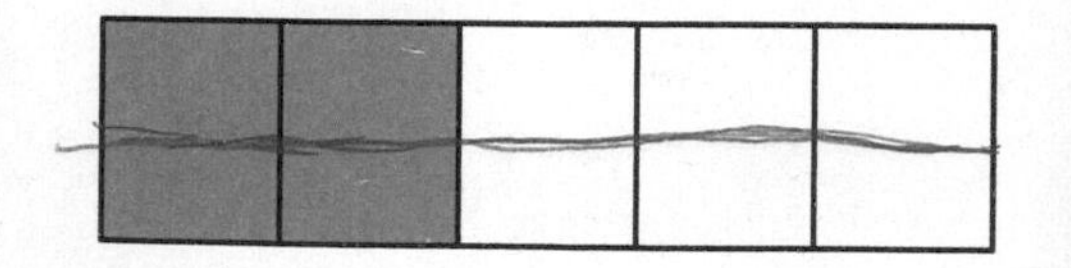

2. Use the number line to find an equivalent fraction for $\frac{1}{6}$.

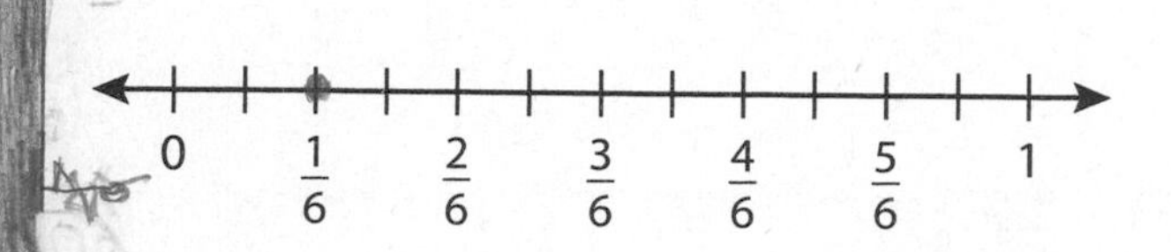

Name ______________________________

Apply and Grow: Practice

Use the model to find an equivalent fraction.

3. $\frac{3}{6}$

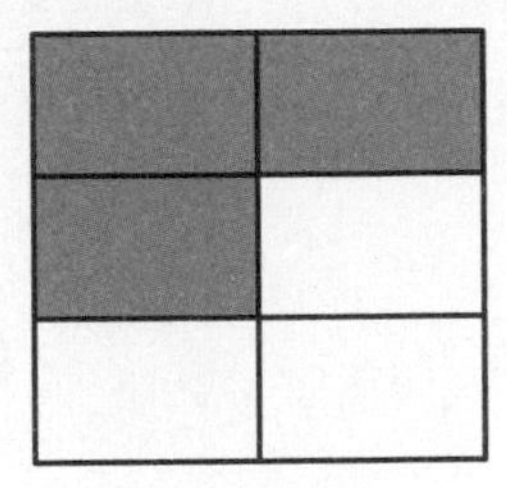

4. $\frac{1}{5}$

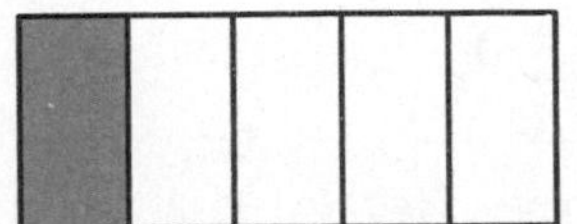

5. $\frac{4}{5}$

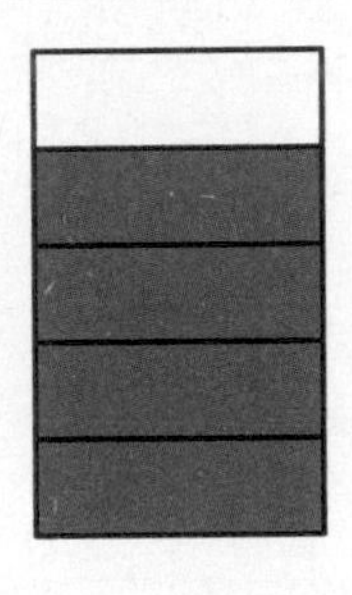

6. $\frac{1}{2}$

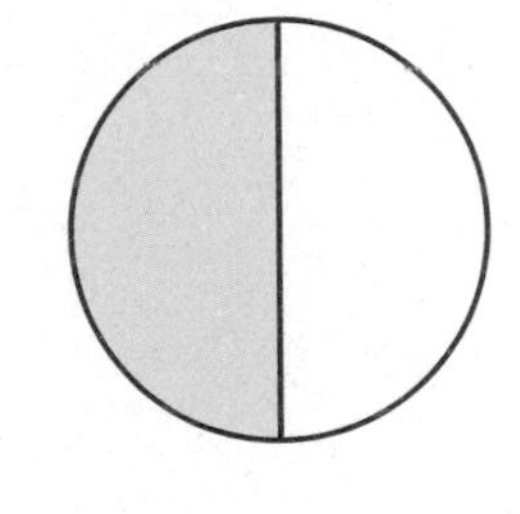

Use the number line to find an equivalent fraction.

7. $\frac{3}{4}$

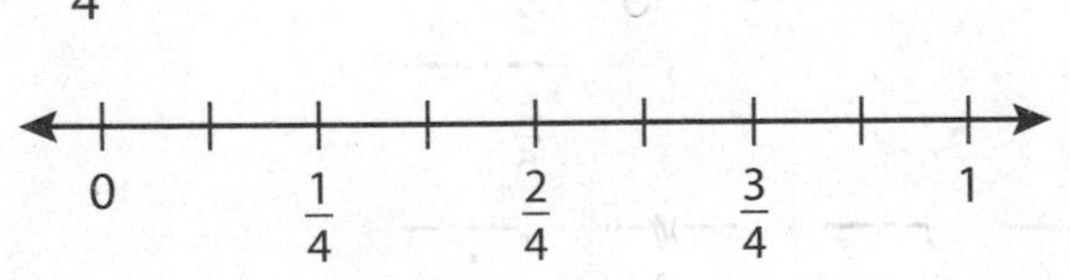

8. $\frac{1}{3}$

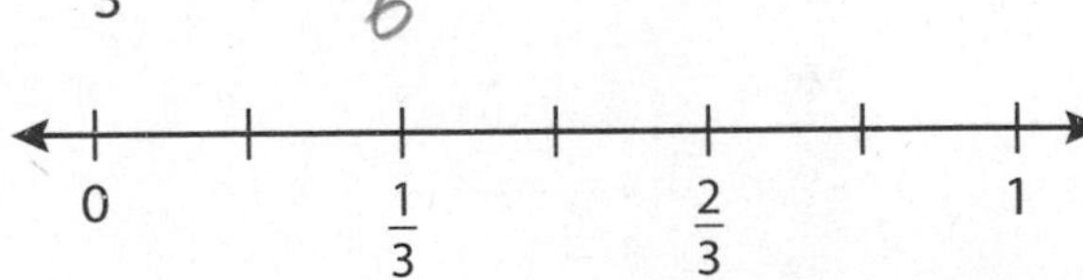

9. Open-Ended Write two equivalent fractions to describe the portion of the eggs that are white.

10. YOU BE THE TEACHER Your friend says the models show equivalent fractions. Is your friend correct? Explain.

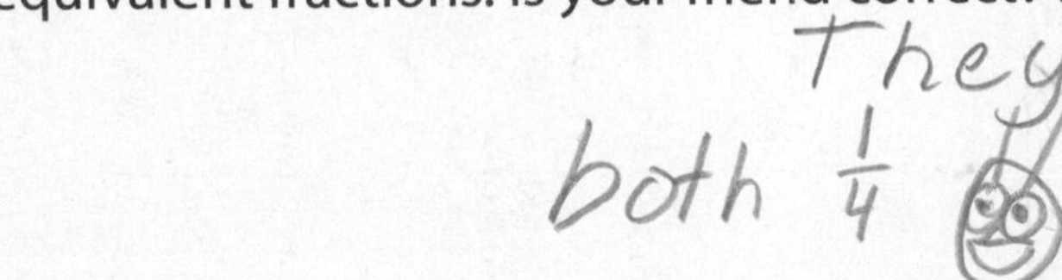

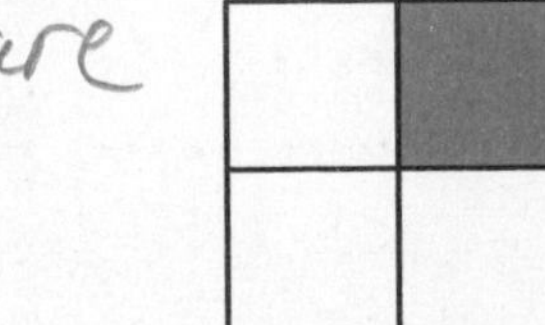

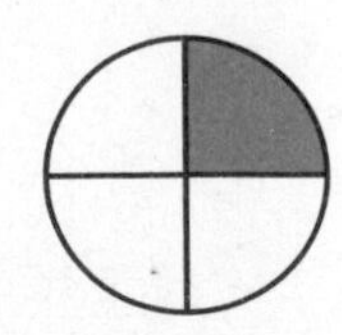

Think and Grow: Modeling Real Life

Example You and your friend make braided paper bookmarks. Yours is $\frac{2}{3}$ foot long. Your friend's is $\frac{7}{12}$ foot long. Are the bookmarks the same length?

Determine whether the fractions are equivalent. Plot the fractions on the same number line. Then compare.

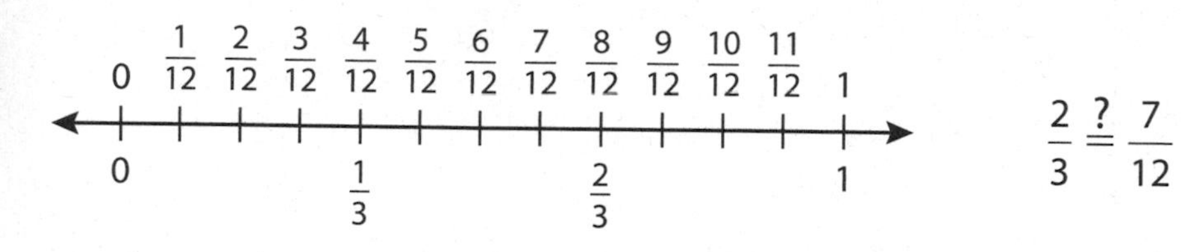

$\frac{2}{3} \stackrel{?}{=} \frac{7}{12}$

So, the bookmarks ________ the same length.

Show and Grow *I can think deeper!*

11. The lasagna pans are the same size. Are the amounts of lasagna left in the pans equal?

$\frac{4}{12}$

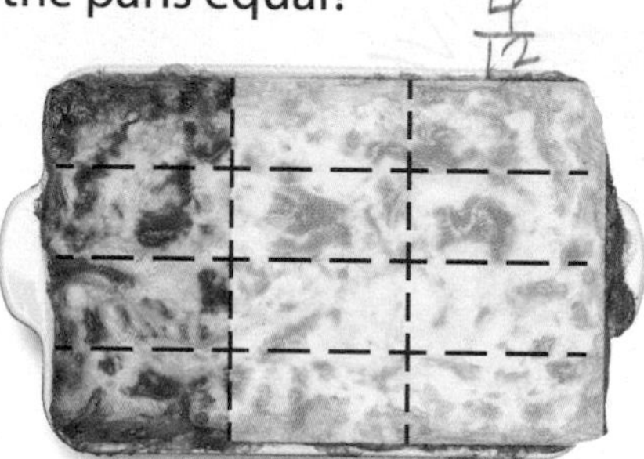

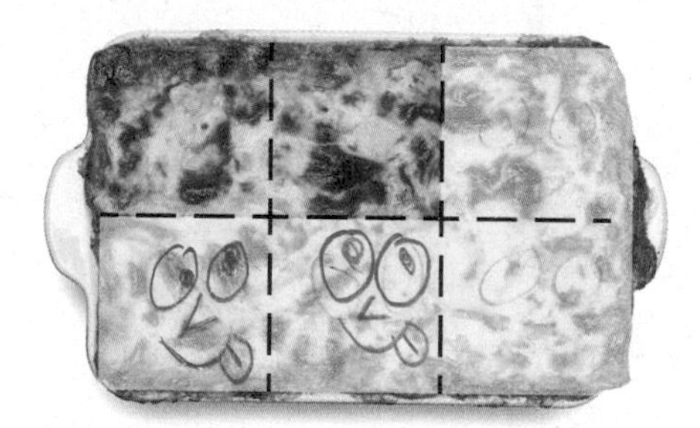

12. **DIG DEEPER!** You run 3 laps around an outdoor track, where 4 laps are equal to 1 mile. Your friend runs 6 laps around the indoor track shown. Do you and your friend run the same distance? Explain.

Both $\frac{3}{4} \times \frac{6}{8}$ (×2, ×2)

8 laps = 1 mile

Name ________________________________

Generate Equivalent Fractions by Multiplying

7.2

Learning Target: Use multiplication to find equivalent fractions.

Success Criteria:

- I can multiply a numerator and a denominator by a chosen number.
- I can multiply to find equivalent fractions.
- I can explain why multiplication can be used to find equivalent fractions.

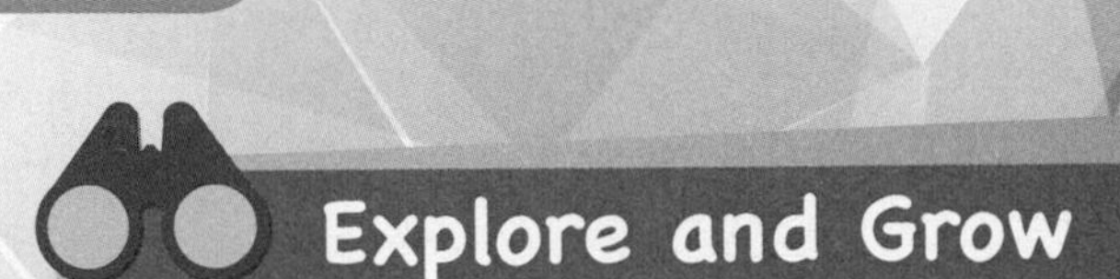

Shade the second model in each pair to show an equivalent fraction. Then write the fraction.

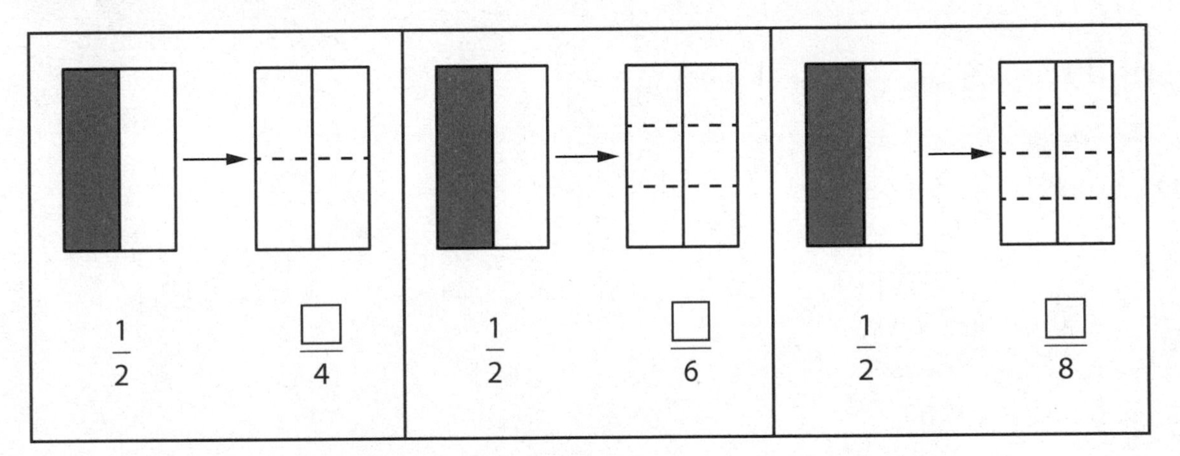

Describe the relationship between each pair of numerators and each pair of denominators.

Structure How can you use multiplication to write equivalent fractions? Explain. Then use your method to find another fraction that is equivalent to $\frac{1}{2}$.

Think and Grow: Multiply to Find Equivalent Fractions

You can find an equivalent fraction by multiplying the numerator and the denominator by the same number.

$$\frac{1}{2} = \frac{1 \times 3}{2 \times 3} = \frac{3}{6}$$

Example Find an equivalent fraction for $\frac{3}{5}$.

Multiply the numerator and the denominator by 2.

$$\frac{3}{5} = \frac{3 \times 2}{5 \times 2} = \frac{\square}{\square}$$

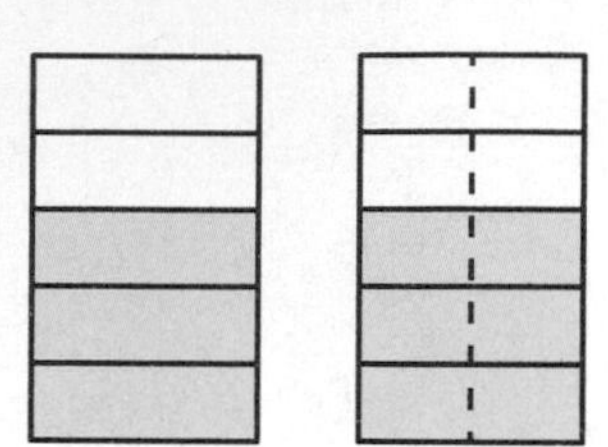

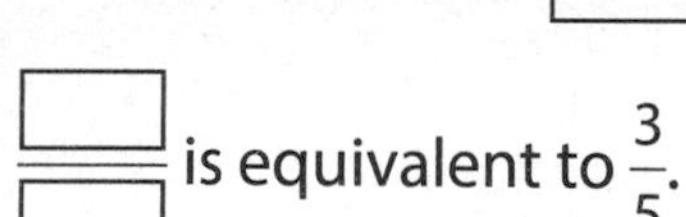

$\frac{\square}{\square}$ is equivalent to $\frac{3}{5}$.

Example Find an equivalent fraction for $\frac{7}{4}$.

Multiply the numerator and the denominator by 3.

$$\frac{7}{4} = \frac{7 \times 3}{4 \times 3} = \frac{\square}{\square}$$

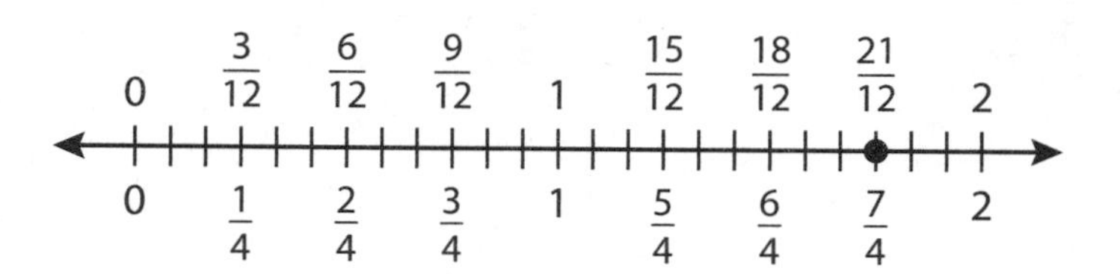

$\frac{\square}{\square}$ is equivalent to $\frac{7}{4}$.

Show and Grow *I can do it!*

Find an equivalent fraction.

1. $\frac{5}{6} = \frac{5 \times \square}{6 \times \square} = \frac{\square}{\square}$

2. $\frac{8}{5} = \frac{8 \times \square}{5 \times \square} = \frac{\square}{\square}$

Find the equivalent fraction.

3. $\frac{1}{2} = \frac{\square}{8}$

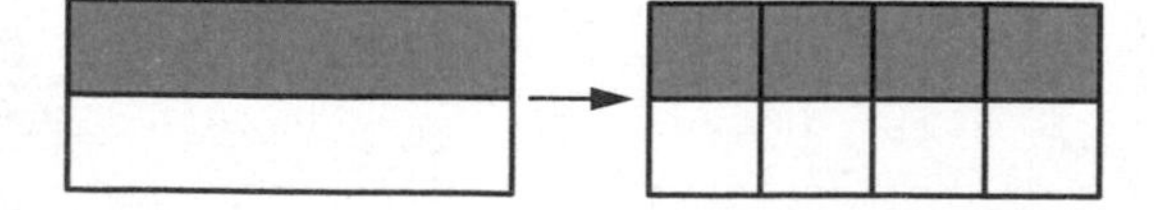

4. $\frac{2}{3} = \frac{\square}{6}$

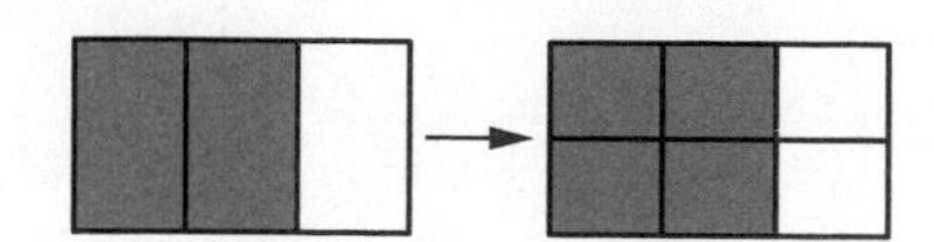

Name ____________________

Apply and Grow: Practice

Find the equivalent fraction.

5. $\frac{3}{4} = \frac{\square}{8}$

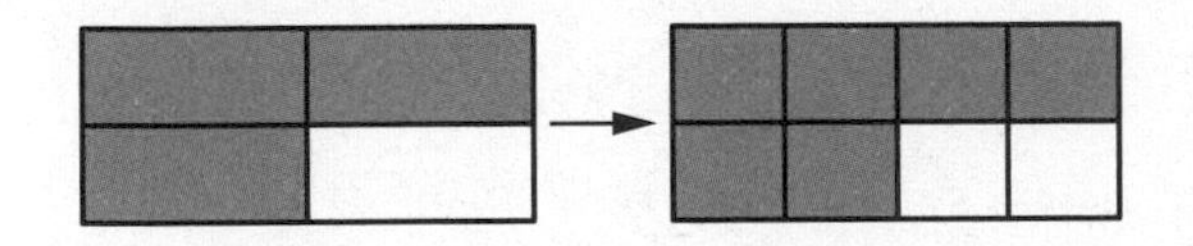

6. $\frac{1}{3} = \frac{4}{\square}$

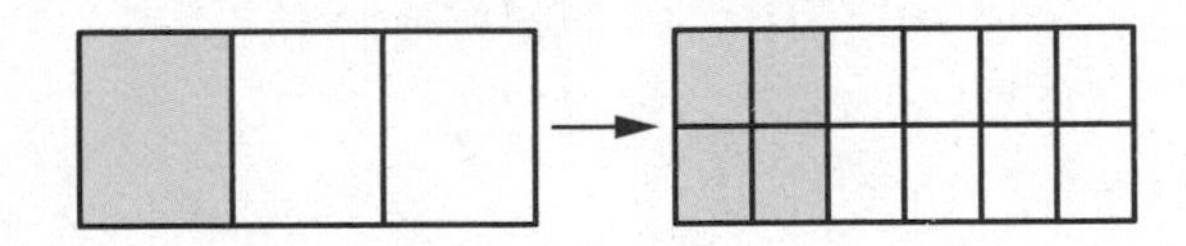

7. $\frac{9}{6} = \frac{18}{\square}$

8. $\frac{7}{5} = \frac{\square}{100}$

Find an equivalent fraction.

9. $\frac{7}{6}$

10. $\frac{10}{10}$

11. $\frac{2}{4}$

Find two equivalent fractions.

12. $\frac{5}{5}$

13. $\frac{4}{3}$

14. $\frac{1}{10}$

15. **Writing** Explain how $\frac{1}{4}$ and $\frac{2}{8}$ are equivalent using multiplication. Use models to support your answer.

DIG DEEPER! Write *true* or *false* for the statement. If false, explain why.

16. $\frac{4}{2} \stackrel{?}{=} \frac{24}{12}$ ________

17. $\frac{3}{5} \stackrel{?}{=} \frac{6}{100}$ ________

Think and Grow: Modeling Real Life

Example A recipe calls for $\frac{3}{4}$ cup of oats. You only have a $\frac{1}{8}$-cup measuring cup. What fraction of a cup of oats, in eighths, do you need? Use multiplication to write an equivalent fraction for $\frac{3}{4}$ in eighths.

$\frac{3}{4} = \frac{3 \times \square}{4 \times \square} = \frac{\square}{8}$ Think: 4 times what number equals 8?

So, you need $\frac{\square}{\square}$ cups of oats.

Show and Grow *I can think deeper!*

18. You need $\frac{1}{2}$ cup of water for a science experiment. You only have a $\frac{1}{4}$-cup measuring cup. What fraction of a cup of water, in fourths, do you need?

19. A pedestrian needs to walk $\frac{4}{5}$ mile to meet her goal. The path is marked at every tenth of a mile. What fraction of a mile, in tenths, should she walk?

20. **DIG DEEPER!** You put together $\frac{7}{10}$ of a puzzle. The puzzle has 100 pieces. What fraction of the puzzle, in hundredths, is *not* put together? Explain.

21. **DIG DEEPER!** You have $\frac{3}{5}$ of a dollar in coins. What fraction of a dollar, in hundredths, do you have? Write one possible combination of coins that you have.

Name ______________________________

Homework & Practice 7.2

Learning Target: Use multiplication to find equivalent fractions.

Example Find an equivalent fraction for $\frac{1}{3}$.

Multiply the numerator and the denominator by 4.

$$\frac{1}{3} = \frac{1 \times \boxed{4}}{3 \times \boxed{4}} = \frac{\boxed{4}}{\boxed{12}}$$

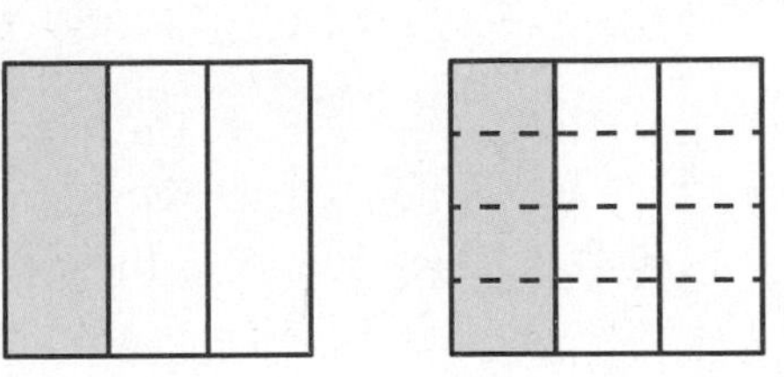

$\frac{\boxed{4}}{\boxed{12}}$ is equivalent to $\frac{1}{3}$.t

Find an equivalent fraction.

1. $\frac{1}{5} = \frac{1 \times \square}{5 \times \square} = \frac{\square}{\square}$

2. $\frac{11}{6} = \frac{11 \times \square}{6 \times \square} = \frac{\square}{\square}$

Find the equivalent fraction.

3. $\frac{4}{6} = \frac{\square}{12}$

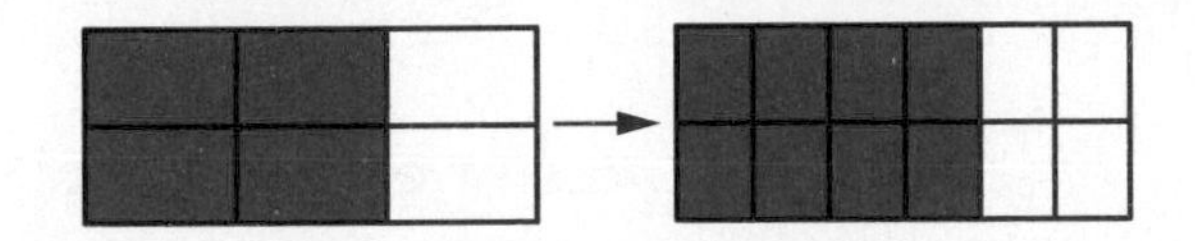

4. $\frac{2}{5} = \frac{4}{\square}$

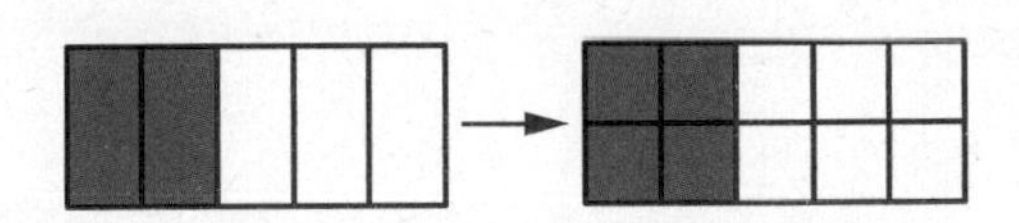

5. $\frac{3}{3} = \frac{\square}{6}$

6. $\frac{7}{10} = \frac{70}{\square}$

Find an equivalent fraction.

7. $\frac{5}{3}$

8. $\frac{4}{4}$

9. $\frac{5}{10}$

Find two equivalent fractions.

10. $\frac{3}{2}$

11. $\frac{4}{10}$

12. $\frac{10}{5}$

13. DIG DEEPER! What is Descartes's fraction?

My fraction is equivalent to $\frac{8}{6}$ and has a numerator that is 4 more than its denominator.

14. YOU BE THE TEACHER Your friend says she can write a fraction equivalent to $\frac{3}{4}$ that has a denominator of 10 and a whole number in the numerator. Is your friend correct? Explain.

15. **Modeling Real Life** A recipe calls for 1 teaspoon of cinnamon. You only have a $\frac{1}{2}$-teaspoon measuring spoon. What fraction of a teaspoon of cinnamon, in halves, do you need?

16. DIG DEEPER! A couple lives in Florida for $\frac{1}{3}$ of the year. Each year has 12 months. What fraction of a year, in twelfths, does the couple *not* live in Florida?

Review & Refresh

Divide. Then check your answer.

17. $7\overline{)891}$

18. $3\overline{)2{,}395}$

19. $6\overline{)627}$

Name ____________________

Generate Equivalent Fractions by Dividing

7.3

Learning Target: Use division to find equivalent fractions.

Success Criteria:

- I can find the factors of a number.
- I can find the common factors of a numerator and a denominator.
- I can divide to find equivalent fractions.

Explore and Grow

Shade the second model in each pair to show an equivalent fraction. Then write the fraction.

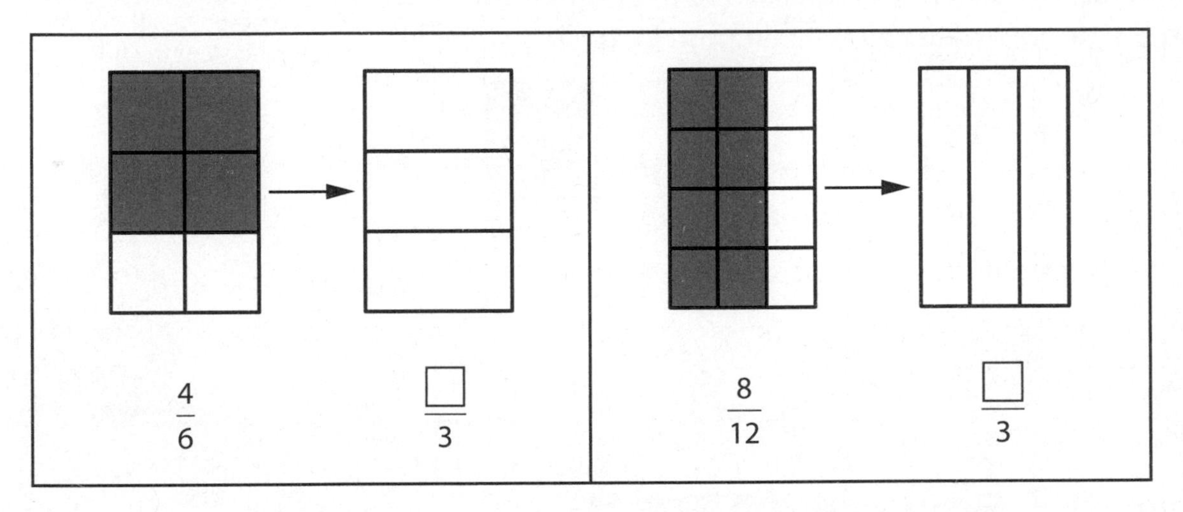

Describe the relationship between each pair of numerators and each pair of denominators.

Structure How can you use division to write equivalent fractions? Explain. Then use your method to find a fraction that is equivalent to $\frac{6}{10}$.

Think and Grow: Divide to Find Equivalent Fractions

A factor that is shared by two or more given numbers is a **common factor**. You can find an equivalent fraction by dividing the numerator and the denominator by a common factor.

$$\frac{2}{4} = \frac{2 \div 2}{4 \div 2} = \frac{1}{2}$$

Example Find an equivalent fraction for $\frac{8}{12}$.

Find the common factors of 8 and 12.

- The factors of 8 are ①, ②, ④, and 8.
- The factors of 12 are ①, ②, 3, ④, 6, and 12.

So, the common factors are ____, ____, and ____.

To find an equivalent fraction, divide the numerator and the denominator by the common factor 2.

$$\frac{8}{12} = \frac{8 \div \square}{12 \div \square} = \frac{\square}{\square}$$

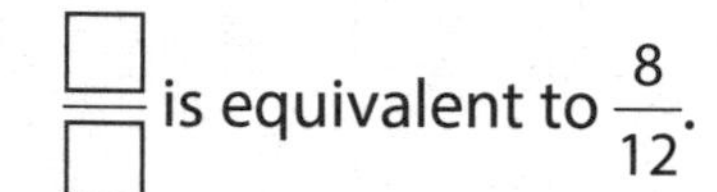

$\frac{\square}{\square}$ is equivalent to $\frac{8}{12}$.

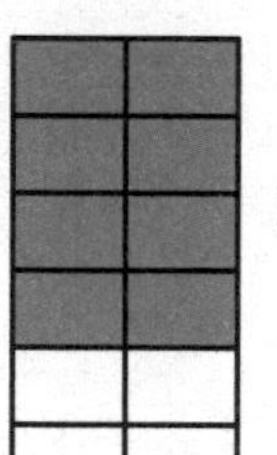

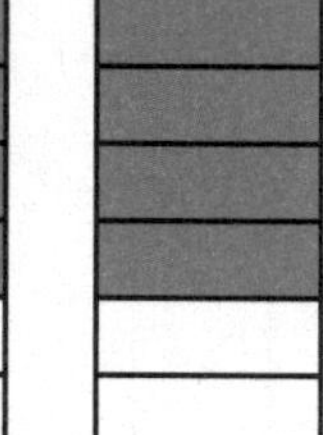

You can also divide the numerator and the denominator by the common factor 4 to find an equivalent fraction.

Show and Grow *I can do it!*

Find an equivalent fraction.

1. $\frac{3}{6} = \frac{3 \div \square}{6 \div \square} = \frac{\square}{\square}$

2. $\frac{20}{8} = \frac{20 \div \square}{8 \div \square} = \frac{\square}{\square}$

Find the equivalent fraction.

3. $\frac{4}{10} = \frac{\square}{5}$

4. $\frac{90}{100} = \frac{9}{\square}$

5. $\frac{14}{4} = \frac{\square}{2}$

Name ______________________________

Apply and Grow: Practice

Find the equivalent fraction.

6. $\frac{2}{6} = \frac{\square}{3}$

7. $\frac{16}{12} = \frac{4}{\square}$

8. $\frac{80}{100} = \frac{\square}{10}$

9. $\frac{8}{8} = \frac{\square}{1}$

10. $\frac{2}{4} = \frac{1}{\square}$

11. $\frac{30}{6} = \frac{\square}{2}$

Find an equivalent fraction for the point on the number line.

12. 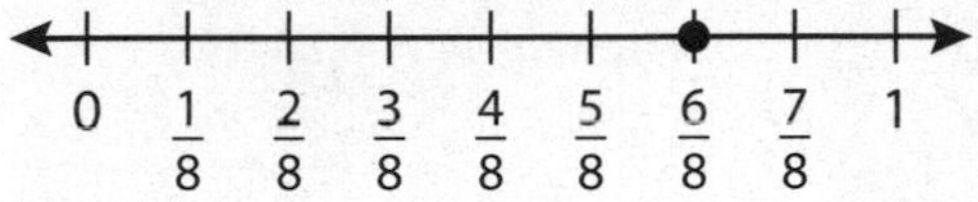

13. 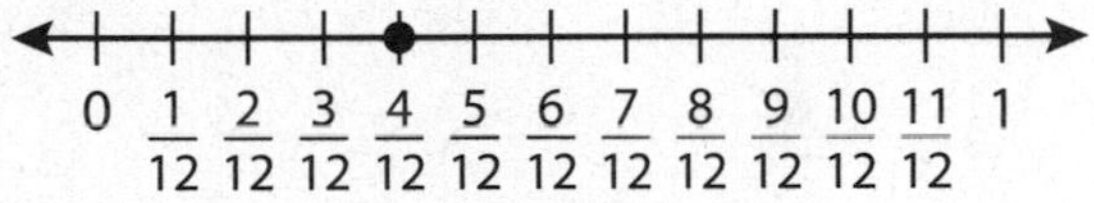

Find an equivalent fraction.

14. $\frac{3}{12}$

15. $\frac{18}{6}$

Find two equivalent fractions.

16. $\frac{20}{10}$

17. $\frac{75}{100}$

18. MP **Reasoning** Your friend begins to divide the numerator and denominator of $\frac{12}{6}$ by 4 and then gets stuck. Explain why your friend gets stuck.

19. **DIG DEEPER!** Can you write an equivalent fraction with a lesser numerator and denominator when the numerator and denominator of a fraction are both odd numbers? Explain.

Think and Grow: Modeling Real Life

Example The Lechtal High Trail is a 100-kilometer hiking trail in Austria. A hiker has completed 70 kilometers of the trail. What fraction of the trail, in tenths, has the hiker completed?

Use the distances to write the fraction of the trail the hiker has completed.

$$\frac{\text{distance hiker has completed}}{\text{total trail length}} = \frac{70}{100}$$

Use division to write an equivalent fraction in tenths.

$$\frac{70}{100} = \frac{70 \div \square}{100 \div \square} = \frac{\square}{10}$$

Think: 100 divided by what number equals 10?

So, the hiker has completed $\frac{\square}{\square}$ of the trail.

Show and Grow *I can think deeper!*

20. A puzzle cube has 54 stickers. Nine of the stickers are orange. A cube has 6 faces. What fraction of the stickers, in sixths, are orange?

21. There are 28 students in a class. Seven of the students pack their lunch. What fraction of the students, in fourths, pack their lunch?

22. **DIG DEEPER!** There are 45 apps on a tablet. Nine of the apps are games. What fraction of the apps, in fifths, are *not* games? Explain.

Name ______________________________

Homework & Practice 7.3

Learning Target: Use division to find equivalent fractions.

Example Find an equivalent fraction for $\frac{4}{8}$.

Find the common factors of 4 and 8.

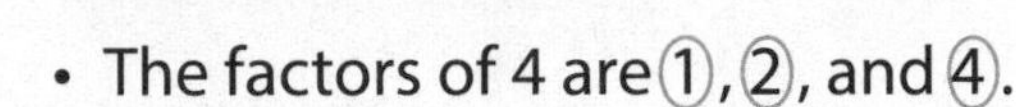

- The factors of 4 are 1, 2, and 4.
- The factors of 8 are 1, 2, 4, and 8.

So, the common factors are 1, 2, and 4.

To find an equivalent fraction, divide the numerator and the denominator by the common factor 4.

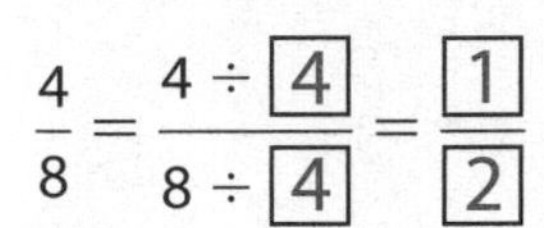

$\frac{4}{8} = \frac{4 \div \boxed{4}}{8 \div \boxed{4}} = \frac{\boxed{1}}{\boxed{2}}$

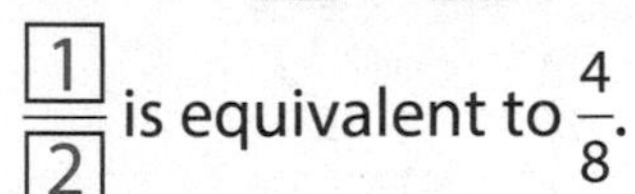

$\frac{\boxed{1}}{\boxed{2}}$ is equivalent to $\frac{4}{8}$.

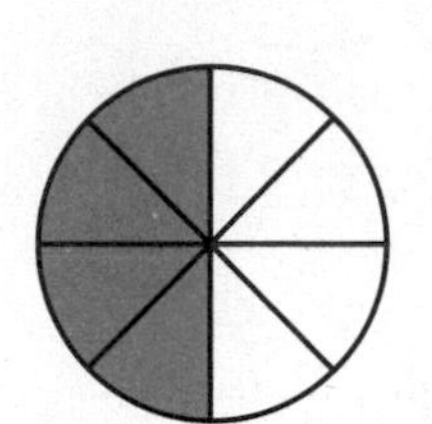

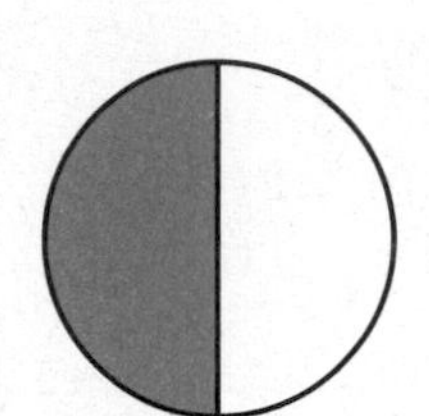

You can also divide the numerator and the denominator by the common factor 2 to find an equivalent fraction.

Find an equivalent fraction.

1. $\frac{8}{10} = \frac{8 \div \square}{10 \div \square} = \frac{\square}{\square}$

2. $\frac{24}{6} = \frac{24 \div \square}{6 \div \square} = \frac{\square}{\square}$

Find the equivalent fraction.

3. $\frac{4}{6} = \frac{\square}{3}$

4. $\frac{25}{100} = \frac{5}{\square}$

Find an equivalent fraction for the point on the number line.

5.

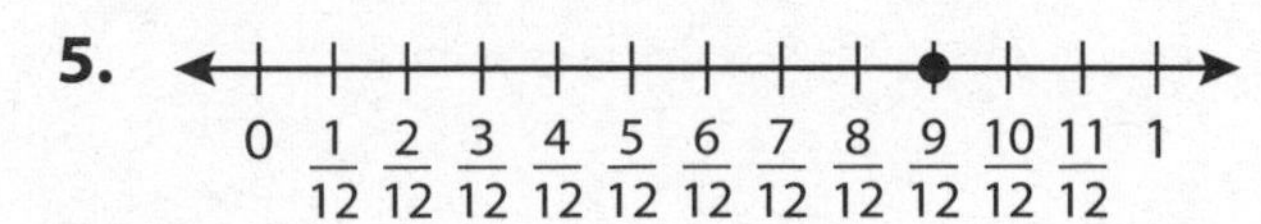

6.

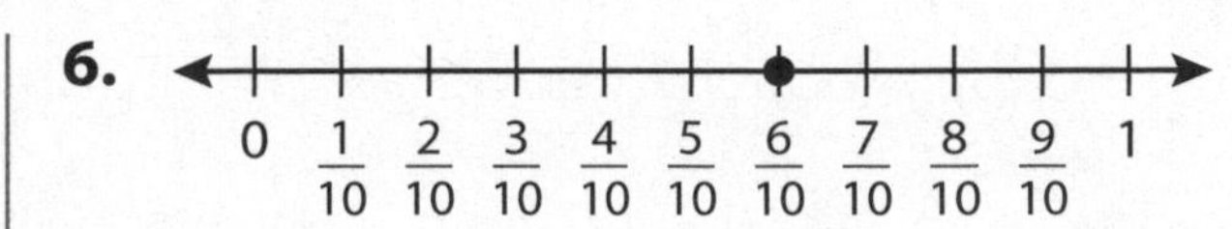

Find an equivalent fraction.

7. $\frac{3}{6}$

8. $\frac{8}{4}$

9. $\frac{15}{5}$

Find two equivalent fractions.

10. $\frac{40}{100}$

11. $\frac{6}{6}$

12. $\frac{24}{8}$

13. **Writing** Explain how to find an equivalent fraction using division.

14. MP **Patterns** Describe and complete the pattern.

$\frac{64}{1{,}600}, \frac{32}{800}, \frac{16}{400}, \frac{\square}{\square}, \frac{\square}{\square}$

15. **Modeling Real Life** A book shows 100 hieroglyphic symbols. You have learned the meanings of 30 of them. What fraction of the symbols' meanings, in tenths, have you learned?

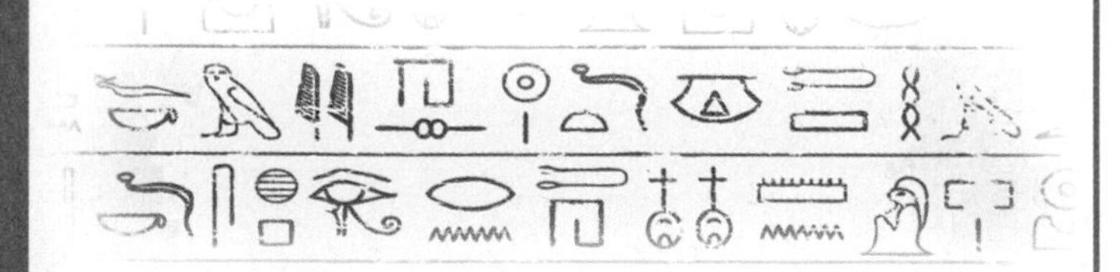

16. **DIG DEEPER!** There are 54 players in a beach volleyball club . Nine of the players cannot attend a game night. The coach needs to make even teams with the players that are there. What fraction of the players, in sixths, are at the game night?

Review & Refresh

Find the product. Check whether your answer is reasonable.

17. $71 \times 63 =$ ______

18. $24 \times 98 =$ ______

19. $85 \times 27 =$ ______

Name ______________________

Compare Fractions Using Benchmarks 7.4

Learning Target: Compare fractions using benchmarks.

Success Criteria:

- I can compare a fraction to a benchmark of $\frac{1}{2}$ or 1.
- I can use a benchmark to compare two fractions.

Use a model to compare $\frac{1}{2}$ and $\frac{5}{8}$.

$\frac{1}{2} \bigcirc \frac{5}{8}$

Use a model to compare $\frac{1}{2}$ and $\frac{2}{5}$.

$\frac{1}{2} \bigcirc \frac{2}{5}$

How can you use your results to compare $\frac{5}{8}$ and $\frac{2}{5}$?

Structure How does the numerator of a fraction compare to the denominator when the fraction is less than $\frac{1}{2}$? greater than $\frac{1}{2}$? equal to $\frac{1}{2}$? Explain.

Think and Grow: Compare Fractions Using Benchmarks

A **benchmark** is a commonly used number that you can use to compare other numbers. You can use the benchmarks $\frac{1}{2}$ and 1 to help you compare fractions.

7 is greater than half of 10, so $\frac{7}{10}$ is greater than $\frac{1}{2}$.

Example Use fraction strips to compare $\frac{7}{10}$ and $\frac{3}{8}$.

Compare each fraction to the benchmark $\frac{1}{2}$.

$\frac{1}{10}$	$\frac{1}{10}$	$\frac{1}{10}$	$\frac{1}{10}$	$\frac{1}{10}$	$\frac{1}{10}$	$\frac{1}{10}$	$\frac{1}{10}$	$\frac{1}{10}$	$\frac{1}{10}$

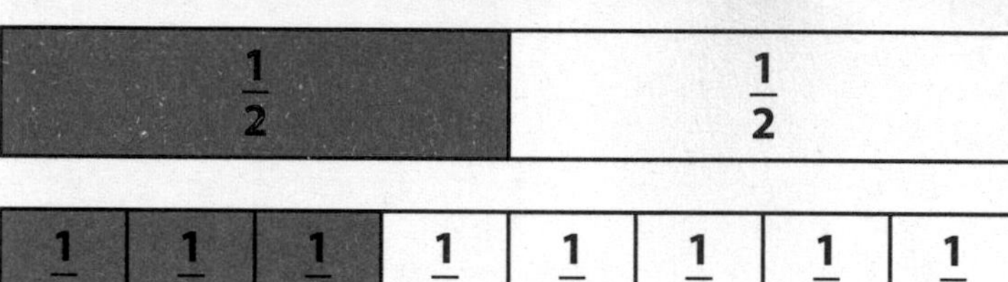

$\frac{1}{8}$	$\frac{1}{8}$	$\frac{1}{8}$	$\frac{1}{8}$	$\frac{1}{8}$	$\frac{1}{8}$	$\frac{1}{8}$	$\frac{1}{8}$

$\frac{7}{10}$ ◯ $\frac{1}{2}$ and $\frac{3}{8}$ ◯ $\frac{1}{2}$

So, $\frac{7}{10}$ (>) $\frac{3}{8}$.

Example Use a number line to compare $\frac{5}{6}$ and $\frac{4}{3}$.

Compare each fraction to the benchmark 1.

$\frac{5}{6}$ ◯ 1 and $\frac{4}{3}$ ◯ 1

So, $\frac{5}{6}$ (<) $\frac{4}{3}$. but ok

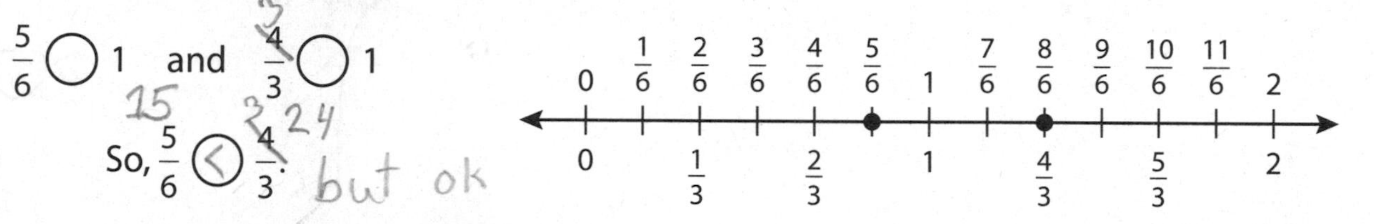

Show and Grow *I can do it!*

Compare. Use a model to help.

1. $\frac{5}{12}$ (<) $\frac{3}{5}$

2. $\frac{3}{4}$ (=) $\frac{6}{8}$

3. $\frac{6}{5}$ (>) $\frac{9}{10}$

Name ______________________________

Apply and Grow: Practice

Compare. Use a model to help.

4. $\frac{4}{12}$ ○ $\frac{7}{10}$

5. $\frac{1}{2}$ ○ $\frac{3}{6}$

6. $\frac{2}{10}$ ○ $\frac{5}{6}$

7. $\frac{5}{5}$ ○ $\frac{12}{12}$

8. $\frac{4}{2}$ ○ $\frac{7}{10}$

9. $\frac{4}{6}$ ○ $\frac{1}{3}$

10. $\frac{5}{4}$ ○ $\frac{3}{8}$

11. $\frac{6}{12}$ ○ $\frac{4}{5}$

12. $\frac{3}{2}$ ○ $\frac{80}{100}$

13. A black bear hibernates for $\frac{7}{12}$ of 1 year. A bat hibernates for $\frac{1}{4}$ of 1 year. Which animal hibernates longer? How do you know?

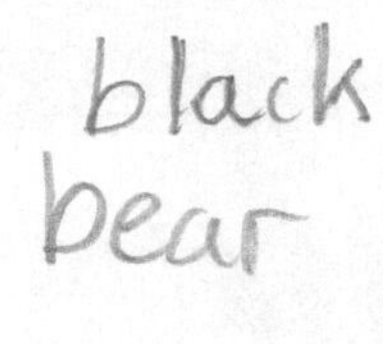

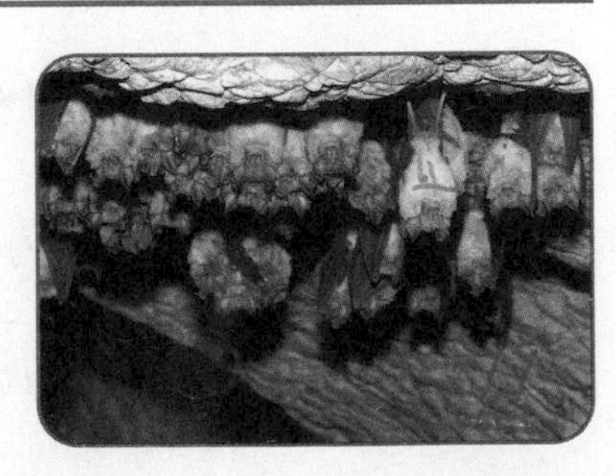

14. Writing Explain how you can tell whether a fraction is greater than, less than, or equal to 1, just by looking at the numerator and the denominator.

15. **DIG DEEPER!** You and your friend pack a lunch. You eat $\frac{2}{6}$ of your lunch. Your friend eats $\frac{3}{4}$ of his lunch. Can you tell who ate more? Explain.

Think and Grow: Modeling Real Life

Example You have $\frac{3}{5}$ of a bottle of blue paint and $\frac{7}{8}$ of a bottle of yellow paint. Do you have enough of each paint color to make the recipe? Explain.

Green Paint Recipe

$\frac{2}{8}$ of a bottle of blue paint

$\frac{5}{10}$ of a bottle of yellow paint

Compare each fraction of a paint bottle to the benchmark $\frac{1}{2}$.

Blue paint: $\frac{3}{5} \bigcirc \frac{1}{2}$ and $\frac{2}{8} \bigcirc \frac{1}{2}$ So, $\frac{3}{5} \bigcirc \frac{2}{8}$.

Yellow paint: $\frac{7}{8} \bigcirc \frac{1}{2}$ and $\frac{5}{10} \bigcirc \frac{1}{2}$ So, $\frac{7}{8} \bigcirc \frac{5}{10}$.

You ________ have enough of each paint color to make the recipe.

Explain.

Show and Grow *I can think deeper!*

16. You have $\frac{3}{8}$ tablespoon of baking soda, $\frac{2}{3}$ tablespoon of contact lens solution, and $\frac{5}{3}$ cups of glue. Do you have enough of each ingredient to make the recipe? Explain.

17. **DIG DEEPER!** You and your friend are making posters for a science fair. The posters are the same size. Your poster has 8 equal parts and you are halfway done. Your friend's poster has 12 equal parts. Your friend has completed $\frac{9}{12}$ of her poster. Who has a greater amount of poster left to complete?

Name ______________________________

Compare Fractions 7.5

Learning Target: Compare fractions using equivalent fractions.

Success Criteria:

- I can compare the numerators and denominators of two fractions.
- I can make the numerators or the denominators of two fractions the same.
- I can compare fractions with like numerators or like denominators.

Explore and Grow

Shade each pair of models to compare $\frac{1}{3}$ and $\frac{5}{12}$. Explain how each pair of models helps you compare the fractions differently.

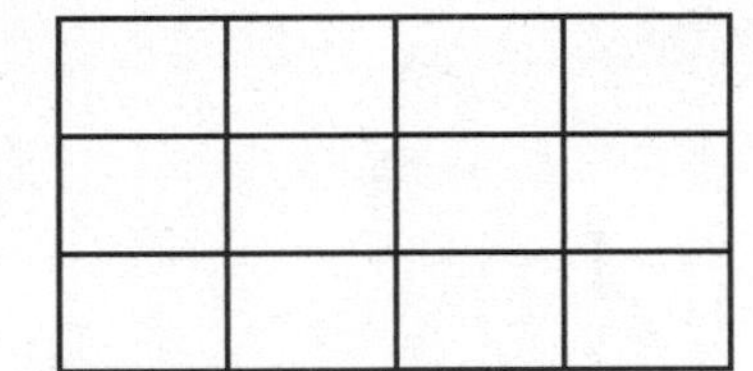

$\frac{1}{3}$

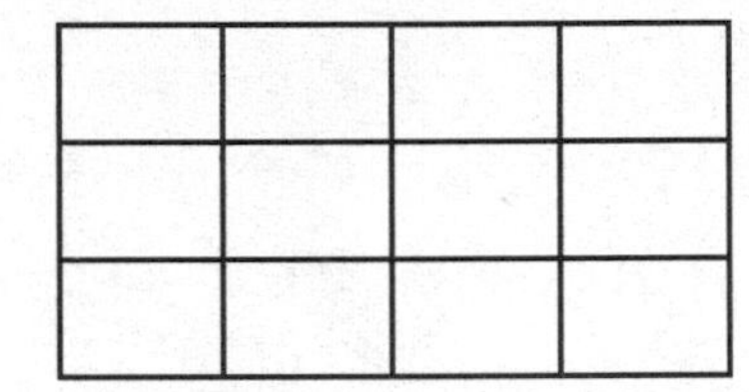

$\frac{5}{12}$

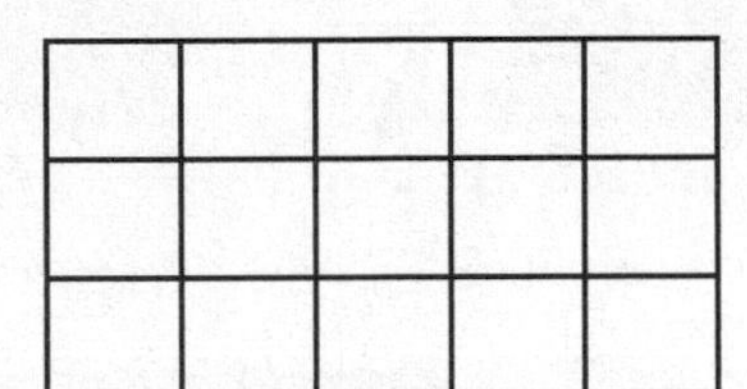

$\frac{1}{3}$

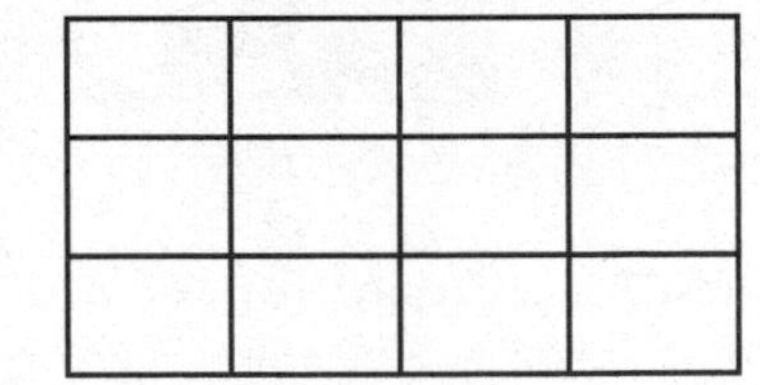

$\frac{5}{12}$

Reasoning How can you use equivalent fractions to compare fractions with different numerators and different denominators?

Here

Think and Grow: Compare Fractions

Example Compare $\frac{3}{5}$ and $\frac{9}{10}$.

One Way: Use a like denominator. Find an equivalent fraction for $\frac{3}{5}$ that has a denominator of 10.

Multiply the numerator and the denominator of $\frac{3}{5}$ by ______.

$\frac{3}{5} = \frac{3 \times \square}{5 \times \square} = \frac{6}{10}$

Compare $\frac{6}{10}$ and $\frac{9}{10}$.

$\frac{6}{10} \bigcirc \frac{9}{10}$

So, $\frac{3}{5} \bigcirc \frac{9}{10}$.

$\frac{3}{5} = \frac{6}{10}$

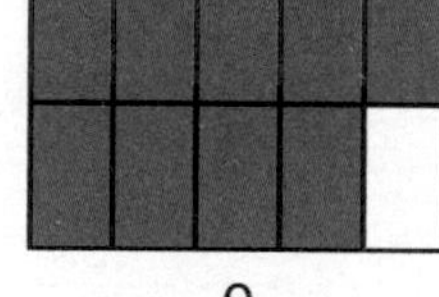

$\frac{9}{10}$

Another Way: Use a like numerator. Find an equivalent fraction for $\frac{3}{5}$ that has a numerator of 9.

Multiply the numerator and the denominator of $\frac{3}{5}$ by ______.

$\frac{3}{5} = \frac{3 \times \square}{5 \times \square} = \frac{9}{15}$

Compare $\frac{9}{15}$ and $\frac{9}{10}$.

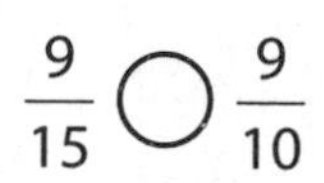

$\frac{9}{15} \bigcirc \frac{9}{10}$

So, $\frac{3}{5} \bigcirc \frac{9}{10}$.

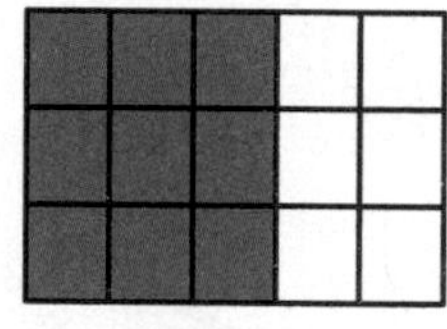

$\frac{3}{5} = \frac{9}{15}$

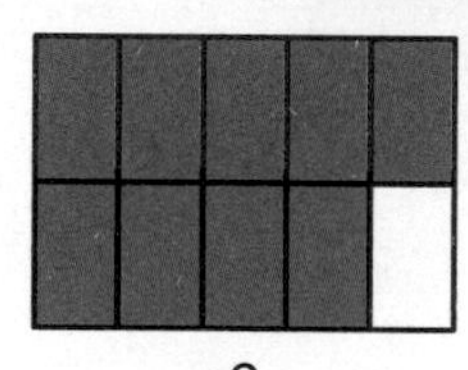

$\frac{9}{10}$

Show and Grow *I can do it!*

Compare. Use a model to help.

1. $\frac{7}{8}$ (>) $\frac{3}{4}$ (handwritten: 28, 24)

2. $\frac{4}{6}$ (=) $\frac{2}{3}$ (handwritten: 12, 12)

3. $\frac{4}{3}$ (>) $\frac{5}{4}$ (handwritten: 16, 15)

Name ______________________________

Apply and Grow: Practice

Compare. Use a model to help.

4. $\frac{6}{12}$ ◯ $\frac{1}{3}$

5. $\frac{8}{10}$ ◯ $\frac{4}{5}$

6. $\frac{3}{10}$ ◯ $\frac{1}{4}$

7. $\frac{3}{8}$ ◯ $\frac{2}{5}$

8. $\frac{9}{6}$ ◯ $\frac{15}{10}$

9. $\frac{7}{10}$ ◯ $\frac{9}{12}$

10. **Writing** Explain why writing fractions with like denominators or like numerators is helpful when comparing them.

11. **DIG DEEPER!** Use the fractions and symbols to make two true statements.

$\frac{9}{12}$	$\frac{5}{6}$	$\frac{7}{8}$	$\frac{2}{3}$	>	<

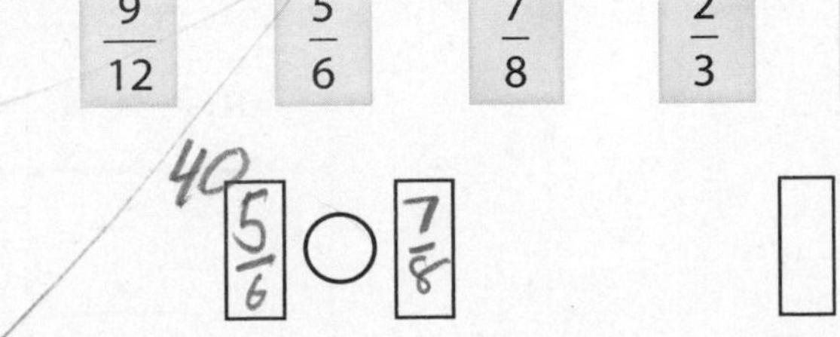

☐ ◯ ☐ ☐ ◯ ☐

Think and Grow: Modeling Real Life

A *socket wrench* is a tool with a metal handle that can be attached to sockets with different sizes.

Example You try to use a $\frac{13}{16}$-inch socket to tighten a bolt, but it is too big. Should you try a $\frac{3}{4}$-inch socket or a $\frac{7}{8}$-inch socket next?

Write each fraction using a like denominator.
Because 4 and 8 are both factors of 16, use 16 as the denominator.

$\frac{3}{4} = \frac{3 \times \square}{4 \times \square} = \frac{\square}{16}$ $\frac{7}{8} = \frac{7 \times \square}{8 \times \square} = \frac{\square}{16}$

Because the $\frac{13}{16}$-inch socket is too big, find the fraction that is less than $\frac{13}{16}$.

$\frac{3}{4} = \frac{\square}{16}$, and $\frac{\square}{16} \bigcirc \frac{13}{16}$. So, $\frac{3}{4} \bigcirc \frac{13}{16}$.

$\frac{7}{8} = \frac{\square}{16}$, and $\frac{\square}{16} \bigcirc \frac{13}{16}$. So, $\frac{7}{8} \bigcirc \frac{13}{16}$.

So, you should try a $\frac{\square}{\square}$-inch socket next.

Show and Grow *I can think deeper!*

12. You drill a hole using a $\frac{5}{16}$-inch drill bit. The hole is too small. Which drill bit should you use to enlarge the hole?

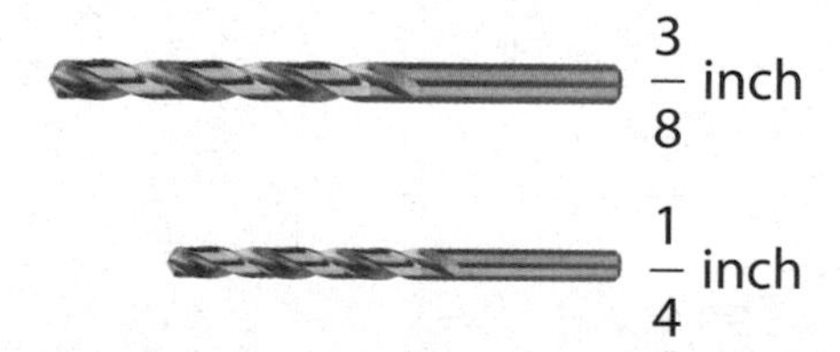

13. **DIG DEEPER!** Order the animals from lightest to heaviest.

Animal	Weight (tons)
Moose	$\frac{4}{5}$
Elk	$\frac{1}{3}$
Grizzly bear	$\frac{3}{8}$

Name ______________________________

Homework & Practice 7.5

Learning Target: Compare fractions using equivalent fractions.

Example Compare $\frac{3}{4}$ and $\frac{6}{8}$.

One Way: Use a like denominator. Find an equivalent fraction for $\frac{3}{4}$ that has a denominator of 8.

Multiply the numerator and the denominator of $\frac{3}{4}$ by __2__.

$\frac{3}{4} = \frac{3 \times \boxed{2}}{4 \times \boxed{2}} = \frac{6}{8}$

Compare $\frac{6}{8}$ and $\frac{6}{8}$.

$\frac{6}{8} \; (=) \; \frac{6}{8}$

So, $\frac{3}{4} \; (=) \; \frac{6}{8}$.

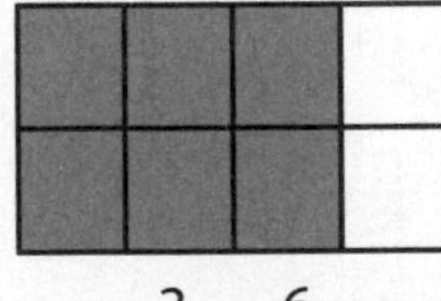

$\frac{3}{4} = \frac{6}{8}$

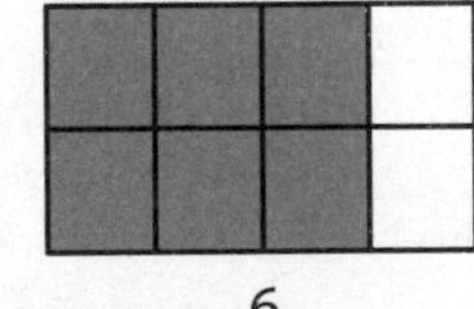

$\frac{6}{8}$

Another Way: Use a like numerator. Find an equivalent fraction for $\frac{6}{8}$ that has a numerator of 3.

Divide the numerator and the denominator of $\frac{6}{8}$ by __2__.

$\frac{6}{8} = \frac{6 \div \boxed{2}}{8 \div \boxed{2}} = \frac{3}{4}$

Compare $\frac{3}{4}$ and $\frac{3}{4}$.

$\frac{3}{4} \; (=) \; \frac{3}{4}$

So, $\frac{6}{8} \; (=) \; \frac{3}{4}$.

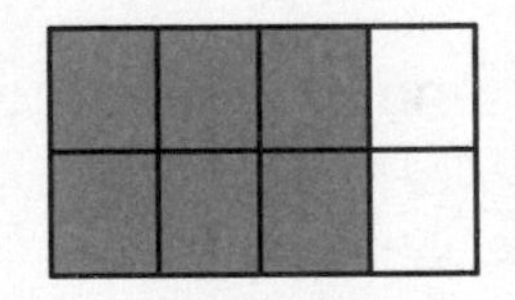

$\frac{6}{8} = \frac{3}{4}$

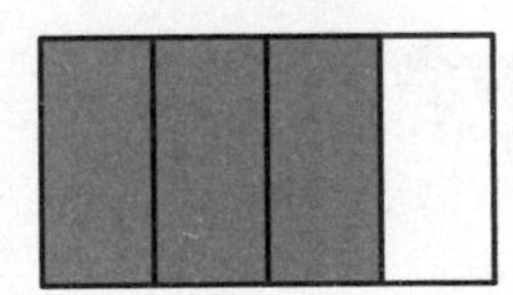

$\frac{3}{4}$

Compare. Use a model to help.

1. $\frac{3}{10} \bigcirc \frac{1}{5}$

2. $\frac{4}{5} \bigcirc \frac{2}{3}$

3. $\frac{5}{8} \bigcirc \frac{2}{1}$

Compare. Use a model to help.

4. $\frac{9}{10}$ ◯ $\frac{97}{100}$

5. $\frac{3}{8}$ ◯ $\frac{2}{6}$

6. $\frac{1}{3}$ ◯ $\frac{4}{12}$

7. $\frac{7}{2}$ ◯ $\frac{6}{5}$

8. $\frac{1}{10}$ ◯ $\frac{2}{12}$

9. $\frac{3}{4}$ ◯ $\frac{4}{6}$

10. MP **Structure** Compare $\frac{3}{8}$ and $\frac{1}{4}$ two different ways.

11. **Modeling Real Life** A sailor is making a ship in a bottle. The last thing he needs to do is seal the bottle with a cork stopper. He tries a $\frac{3}{4}$-inch cork stopper, but it is too small. Should he try a $\frac{1}{2}$-inch cork stopper or a $\frac{4}{5}$-inch cork stopper next? Explain.

12. **DIG DEEPER!** Order the lengths of hair donated from greatest to least.

Student	Hair Lengths Donated (feet)
Student A	$\frac{3}{4}$
Student B	$\frac{11}{12}$
Student C	$\frac{5}{6}$

Review & Refresh

13. Extend the pattern of shapes by repeating the rule "triangle, pentagon, octagon." What is the 48th shape in the pattern?

____ ____ ____ ____ ____ ____ . . .

Name ______________________________

Performance Task 7

1. a. Your art teacher wants you to complete the design below. Half of the squares are colored black. Complete the table. Then use the table to finish the design.

Color	Number of Squares	Fraction of Total Squares
Black		$\frac{1}{2}$
Red	10	
Orange		$\frac{1}{20}$
Yellow		$\frac{2}{25}$
Green	7	
Blue	8	
White		$\frac{3}{25}$

b. Which two colors cover the same portion of the design? Explain.

2. Your teacher displays 30 designs in a rectangular array on the wall. Show two different ways your teacher can arrange the designs.

Fraction Boss

Directions:

1. Divide the Fraction Boss Cards equally between both players.
2. Each player flips a Fraction Boss Card.
3. Players compare their fractions. The player with the greater fraction takes both cards.
4. If the fractions are equal, each player flips another card. Players compare their fractions. The player with the greater fraction takes all four cards.
5. The player with the most cards at the end of the round wins!

Player A	Player B

Name ________________________________

Cumulative Practice 1–7

1. Which number is a common factor of 12, 16, and 40?

Ⓐ 5

Ⓑ 8

Ⓒ 3

Ⓓ 4

---

2. Which statements describe the difference of 77,986 and 21,403?

☐ The difference is greater than 60,000.

☐ The difference is about 60,000.

☐ The difference is less than 60,000.

☐ The difference is 56,583.

---

3. Which product is between 5,050 and 5,100?

Ⓐ 652×8

Ⓑ 566×9

Ⓒ $1{,}023 \times 5$

Ⓓ $1{,}257 \times 4$

---

4. Use the advertisement to answer the question.

What is the greatest number of miles that the car could have been driven?

Ⓐ 84,699

Ⓑ 84,750

Ⓒ 84,749

Ⓓ 84,650

5. Which multiplication expressions could be represented by the area model?

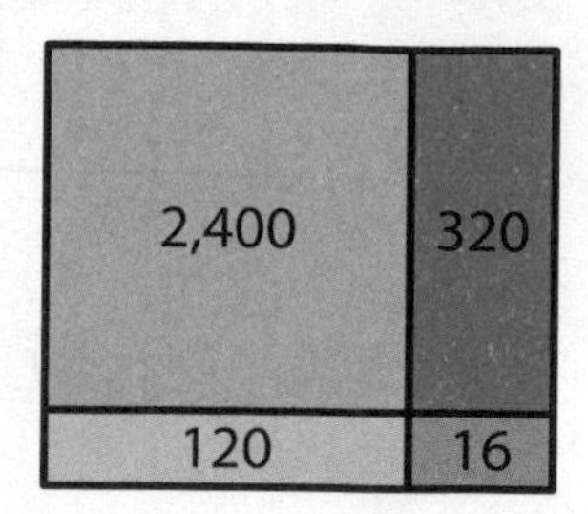

☐ 84×34

☐ 42×68

☐ 240×10

☐ 20×6

---

6. Your neighbor buys yogurt in packages of 4. If your neighbor only buys complete packages, how many yogurts could he have bought?

Ⓐ 2

Ⓑ 20

Ⓒ 34

Ⓓ 18

---

7. Which fraction makes the statement true?

$$? < \frac{6}{8}$$

Ⓐ $\frac{1}{4}$

Ⓑ $\frac{7}{8}$

Ⓒ $\frac{6}{6}$

Ⓓ $\frac{3}{4}$

---

8. Which division equation is represented by the counters?

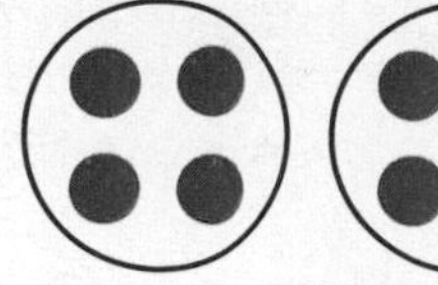 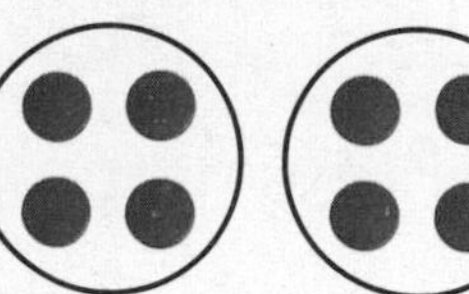 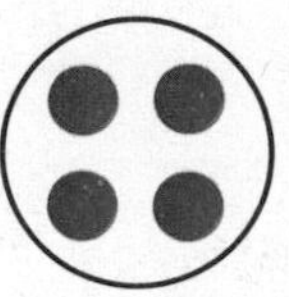 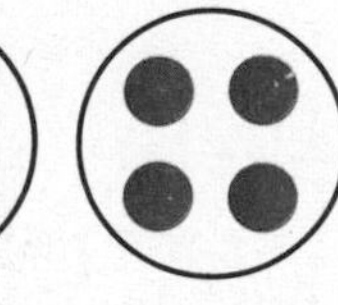 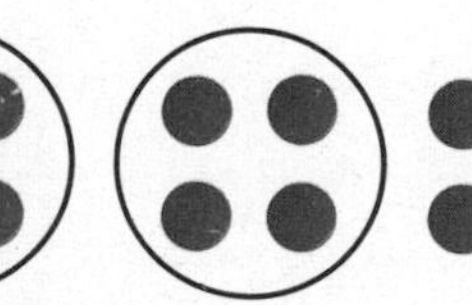

Ⓐ $6 \div 4 \stackrel{?}{=} 2$

Ⓑ $26 \div 6 \stackrel{?}{=} 2 \text{ R}4$

Ⓒ $26 \div 6 \stackrel{?}{=} 4$

Ⓓ $26 \div 6 \stackrel{?}{=} 4 \text{ R}2$

9. Which statements are true?

- ◯ All prime numbers are odd.
- ◯ A composite number cannot have 3 factors.
- ◯ 99 is a prime number.
- ◯ A prime number's factors are 1 and itself.
- ◯ 27 is a composite number.
- ◯ All even numbers greater than 1 are composite.

---

10. What is the quotient of 3,258 and 3?

---

11. Which model shows an equivalent fraction for the fraction shown by the model below?

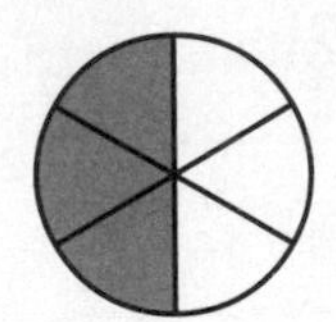

Ⓐ

Ⓑ 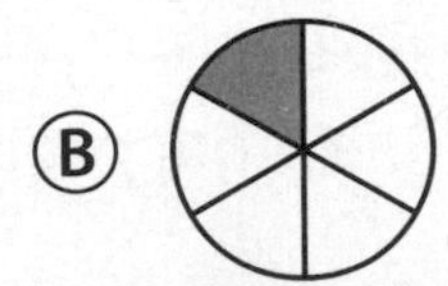

Ⓒ

Ⓓ

---

12. What is the product of 68 and 45?

Ⓐ 8,420

Ⓑ 3,060

Ⓒ 612

Ⓓ 27,360

---

13. Which pattern uses the same rule as the pattern below?

2, 10, 18, 26, 34, 42

Ⓐ 15, 23, 31, 39, 47, 55

Ⓑ 5, 25, 125, 625, 3,125, 15,625

Ⓒ 70, 62, 54, 46, 38, 30

Ⓓ 25, 34, 43, 52, 61, 70

14. Which statements are true?

○ $\frac{1}{5} \stackrel{?}{=} \frac{2}{10}$ ○ $\frac{1}{3} \stackrel{?}{=} \frac{3}{9}$ ○ $\frac{1}{4} \stackrel{?}{=} \frac{1}{8}$

○ $\frac{2}{8} \stackrel{?}{=} \frac{1}{4}$ ○ $\frac{4}{6} \stackrel{?}{=} \frac{2}{4}$ ○ $\frac{6}{12} \stackrel{?}{=} \frac{1}{2}$

---

15. Think Solve Explain

Part A At a summer camp, 67 students are in a line to rent kayaks. Each kayak can hold 4 people. How many kayaks will be full?

Part B How many kayaks will be used?

Part C How many students will be in the last kayak? Explain.

---

16. How many zeros will the product of 80 and 50 have?

Ⓐ 1 Ⓑ 2

Ⓒ 3 Ⓓ 4

---

17. There are 57 electronic books checked out of a library. There are 8 times as many printed books checked out as electronic books. How many total books are checked out of the library?

Ⓐ 513 Ⓑ 4,113

Ⓒ 122 Ⓓ 456

---

18. Which number is equal to 100,000 + 5,000 + 80 + 4?

Ⓐ 1,584 Ⓑ 105,084

Ⓒ 15,084 Ⓓ 105,840

Name ____________________

STEAM Performance Task 1–7

1. Sea level is the average level of the oceans on Earth. The global sea level is rising about $\frac{1}{8}$ inch each year.

   a. If this pattern continues, about how much will the sea level rise in 80 years?

   b. You read from another source that the sea level is rising about $\frac{1}{2}$ inch every 4 years. Did your source use the same fact as above? Explain.

   c. In 10 years, will the sea level rise more or less than an inch? Explain.

   d. How many years will it take the sea level to rise about 1 foot?

Remember, there are 12 inches in one foot.

   e. Use the Internet or some other resource to learn about rising sea levels. Write one interesting fact that you learn.

2. The gravitational pull of the moon affects the high and low tides of the oceans on Earth. Cities along the coasts use tide tables each day. Use the tide table to answer the questions.

Tide Table	
Time of Day	**Height of Water**
7:00 A.M.	48 in.
10:00 A.M.	28 in.
1:00 P.M.	8 in.
4:00 P.M.	32 in.
7:00 P.M.	56 in.
10:00 P.M.	32 in.
1:00 A.M.	4 in.

a. When swimming, why is it important to understand tide tables?

b. What are the common factors of the water heights?

c. Make a picture graph of the water heights.

Each ○ = ______ inches.

d. What pattern do you notice about the water heights? Explain.

Glossary

acute angle **[ángulo agudo]**

An angle that is open less than a right angle

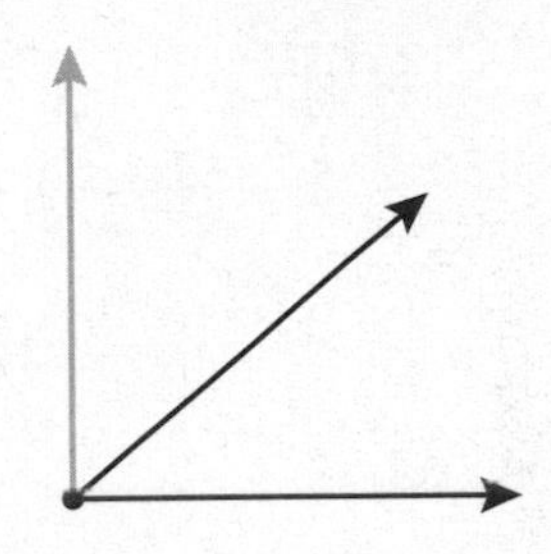

acute triangle **[triángulo acutángulo]**

A triangle that has three acute angles

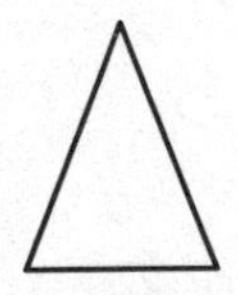

adjacent angles **[ángulos adyacentes]**

Two angles that share a common side and a common vertex, but have no other points in common

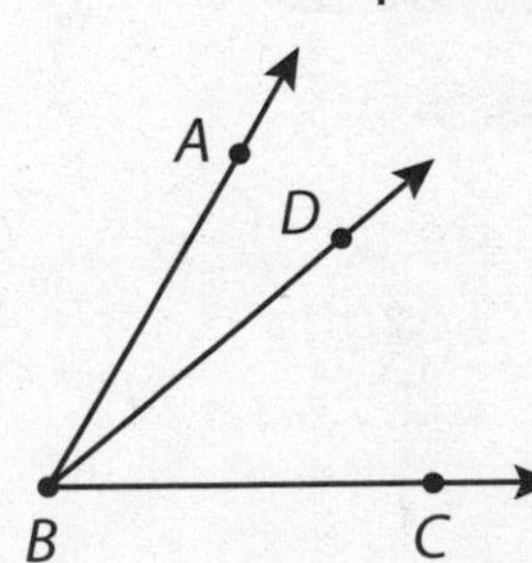

$\angle ABD$ and $\angle DBC$ are adjacent angles.

angle **[ángulo]**

Two rays or line segments that have a common endpoint

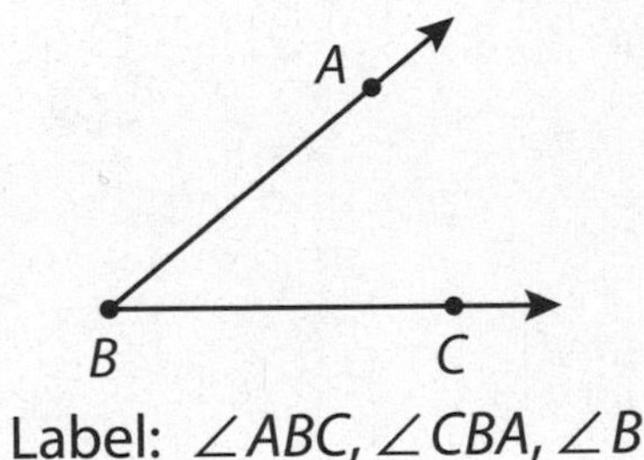

Label: $\angle ABC$, $\angle CBA$, $\angle B$

area **[área]**

The amount of surface a figure covers

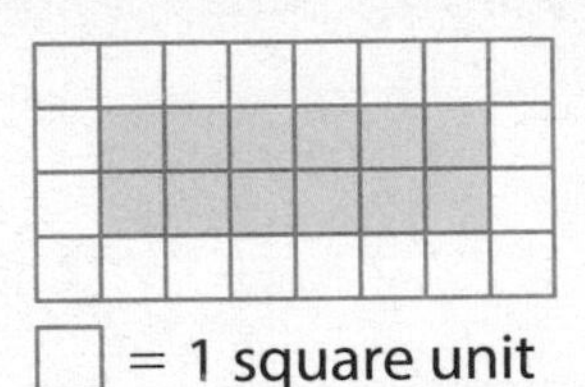

☐ = 1 square unit

The area of the rectangle is 12 square units.

B

benchmark **[punto de referencia]**

A commonly used number that you can use to compare other numbers

Examples: $\frac{1}{2}$, 1

common factor **[factor común]**

A factor that is shared by two or more given numbers

Factors of 8: ①, ②, ④, 8

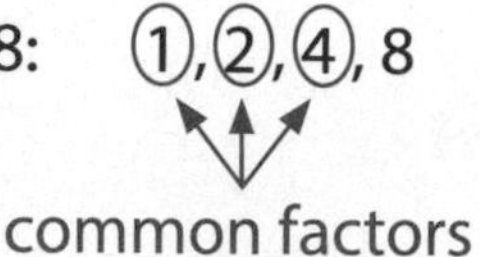

common factors

Factors of 12: ①, ②, 3, ④, 6, 12

---

compatible numbers
[números compatibles]

Numbers that are easy to multiply and are close to the actual numbers

24×31

↓ ↓

25×30

---

complementary angles
[ángulos complementarios]

Two angles whose measures have a sum of 90°

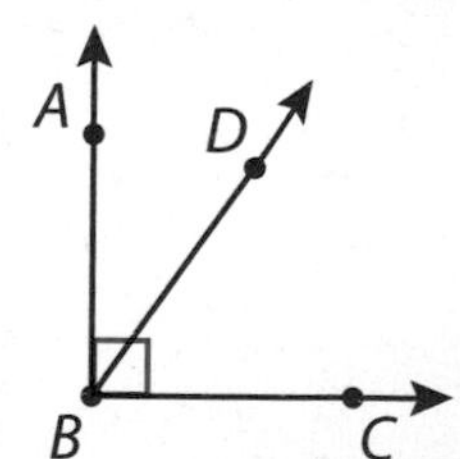

$\angle ABD$ and $\angle DBC$ are complementary angles.

---

composite number
[número compuesto]

A whole number greater than 1 with more than two factors

27

The factors of 27 are 1, 3, 9, and 27.

---

cup (c) **[taza (tz)]**

A customary unit used to measure capacity

The capacity of the measuring cup is 1 cup.

decimal **[decimal]**

A number with one or more digits to the right of the decimal point

0.3

0.04

0.59

---

decimal fraction **[fracción decimal]**

A fraction with a denominator of 10 or 100

$\frac{26}{100}$

$\frac{9}{10}$

$\frac{60}{100}$

---

decimal point **[punto decimal]**

A symbol used to separate the ones place and the tenths place in numbers, and to separate the whole dollars and the cents in money

0.1 $5.06

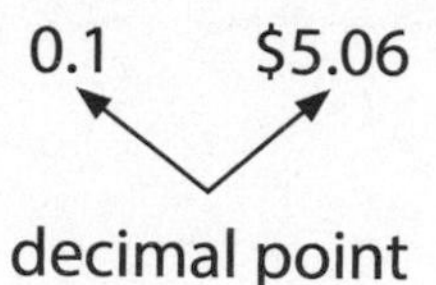

decimal point

degree (°) **[grado (°)]**

The unit used to measure angles

$1° = \frac{1}{360}$ of a circle

Distributive Property
[propiedad distributiva]

$3 \times (5 + 2) = (3 \times 5) + (3 \times 2)$

$3 \times (5 - 2) = (3 \times 5) - (3 \times 2)$

divisible **[divisible]**

A number is divisible by another number when the quotient is a whole number and the remainder is 0.

$48 \div 4 = 12 \text{ R}0$

So, 48 is divisible by 4.

endpoints **[puntos extremos]**

Points that represent the ends of a line segment or ray

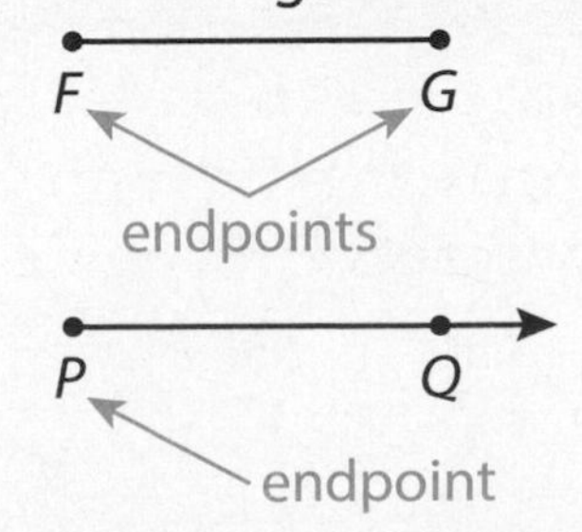

equiangular triangle
[triángulo equiángulo]

A triangle that has three angles with the same measure

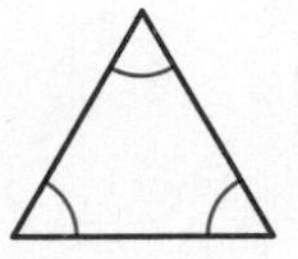

equilateral triangle
[triángulo equilátero]

A triangle that has three sides with the same length

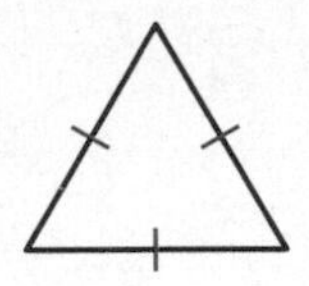

equivalent **[equivalente]**

Having the same value

$\frac{8}{8} = 1$

$3 = \frac{3}{1}$

$2 = \frac{4}{2} = \frac{6}{3}$

equivalent decimals
[decimales equivalente]

Two or more decimals that have the same value

$0.40 = 0.4$

equivalent fractions
[fracciones equivalentes]

Two or more fractions that name the same part of a whole

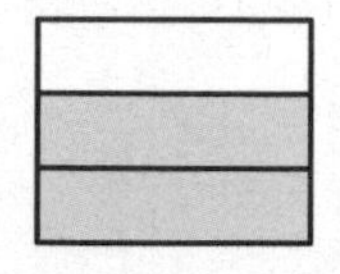
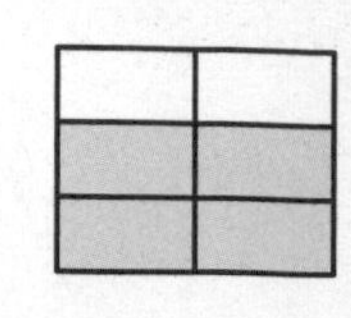

$\frac{2}{3} = \frac{4}{6}$

estimate **[estimación]**

A number that is close to an exact number

$8{,}195 + 9{,}726 = ?$

factor pair **[par de factores]**

Two factors that, when multiplied, result in a given product

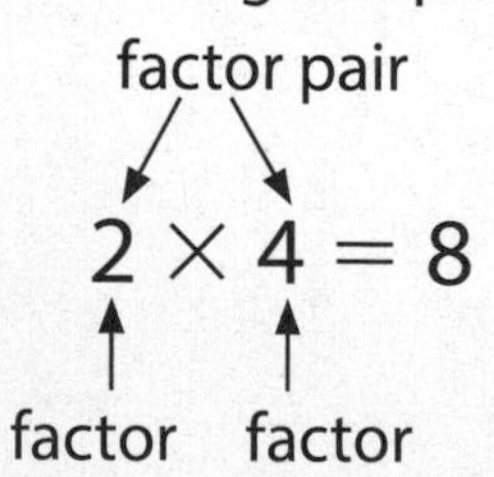

2 and 4 are a factor pair for 8.

formula **[fórmula]**

An equation that uses letters and numbers to show how quantities are related

$P = (2 \times \ell) + (2 \times w)$

$A = \ell \times w$

gallon (gal) **[galón (gal)]**

A customary unit used to measure capacity
There are 4 quarts in 1 gallon.

The capacity of the jug is 1 gallon.

hundredth **[centésimo]**

1 of 100 equal parts of a whole

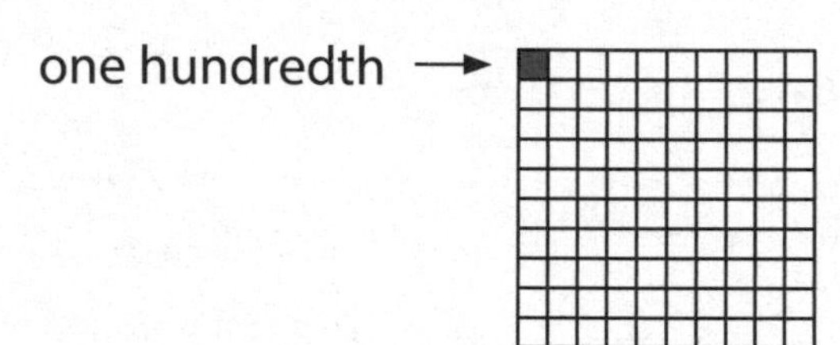

hundredths place
[posición de los centésimos]

The second place to the right of the decimal point

0.01
↑
hundredths place

intersecting lines **[líneas secantes]**

Lines that cross at exactly one point

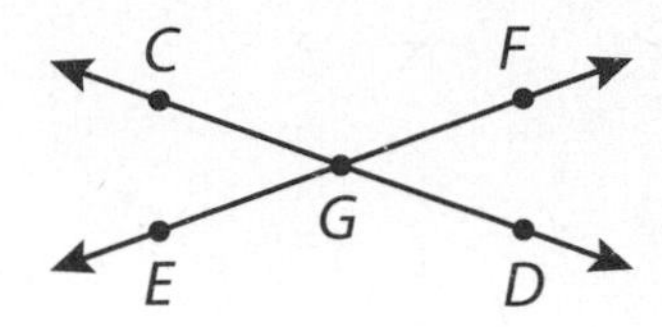

isosceles triangle **[triángulo isósceles]**

A triangle that has two sides with the same length

kilometer (km) **[kilómetro (km)]**

A metric unit used to measure length
There are 1,000 meters in 1 kilometer.

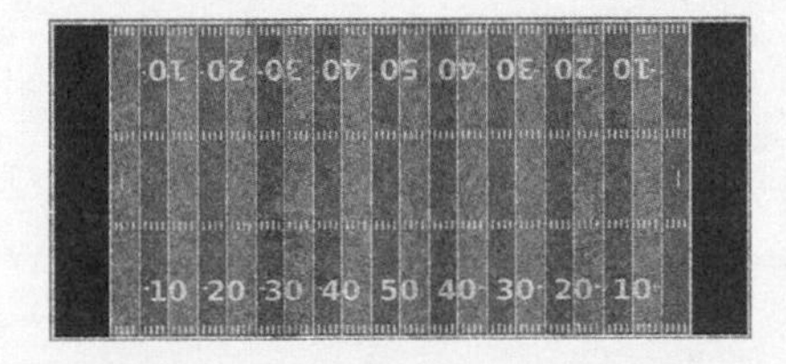

1 kilometer is about the length of 10 football fields including the end zones.

line **[línea]**

A straight path of points that goes on without end in both directions

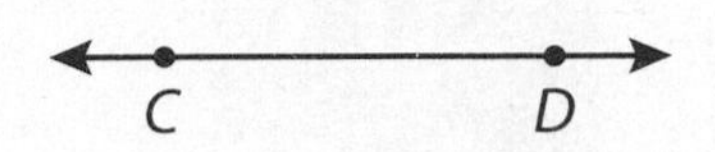

Label: $\overleftrightarrow{CD}$, $\overleftrightarrow{DC}$

line of symmetry **[línea de simetría]**

A fold line that divides a shape into two parts that match exactly

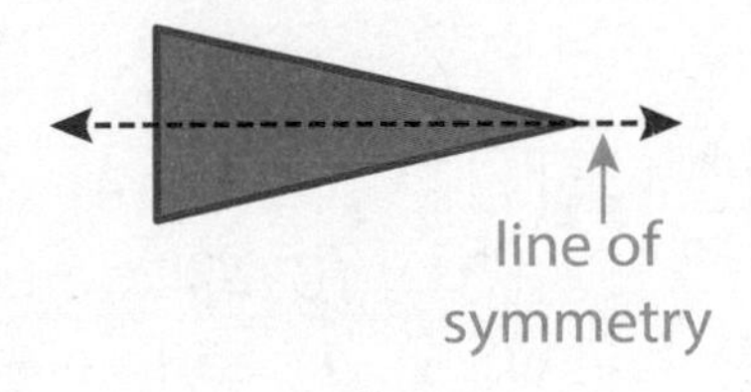

line segment **[segmento lineal]**

A part of a line that includes two endpoints and all of the points between them

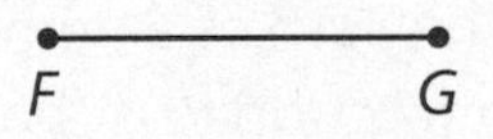

Label: $\overline{FG}$, $\overline{GF}$

line symmetry **[simetría lineal]**

The symmetry that a shape has when it can be folded on a line so that two parts match exactly

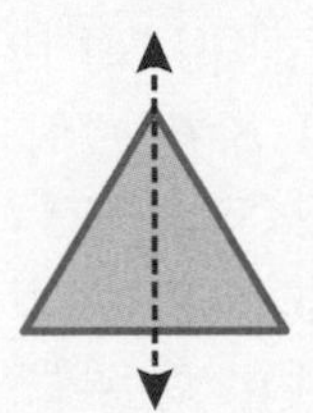

mile (mi) **[milla (mi)]**

A customary unit used to measure length
There are 1,760 yards in 1 mile.

When walking briskly, you can walk
1 mile in about 20 minutes.

millimeter (mm) **[milímetro (mm)]**

A metric unit used to measure length

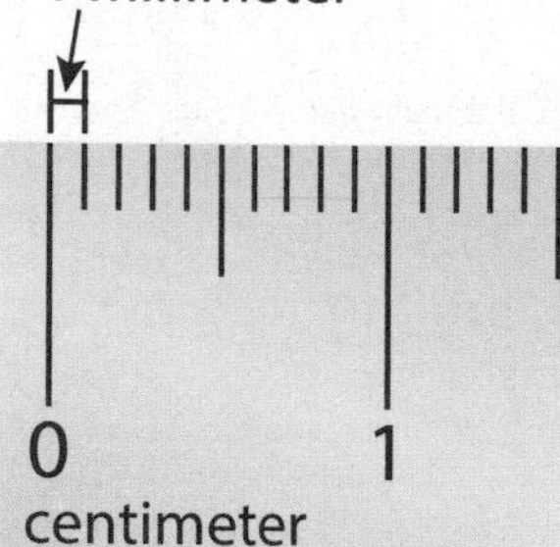

There are
10 millimeters in
1 centimeter.

mixed number **[número mixto]**

Represents the sum of a whole number
and a fraction less than 1

Examples: $2\frac{1}{3}$, $1\frac{4}{5}$, $5\frac{3}{10}$

multiple **[múltiplo]**

The product of a number and any other
counting number

$1 \times 4 = 4$
$2 \times 4 = 8$
$3 \times 4 = 12$
$4 \times 4 = 16$

multiples of 4

obtuse angle **[ángulo obtuso]**

An angle that is open more than a right
angle and less than a straight angle

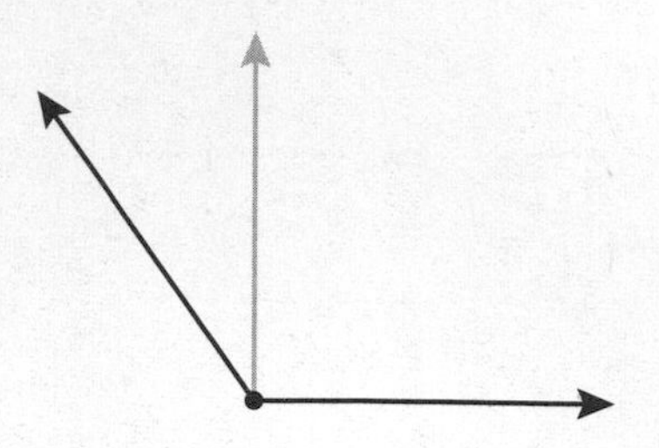

obtuse triangle
[tríangulo obtusángulo]

A triangle that has one obtuse angle

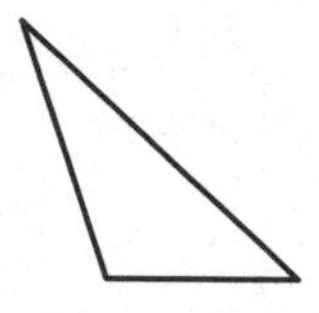

ones period **[período de las unidades]**

The first period in a number

Thousands Period			Ones Period		
Hundreds	Tens	Ones	Hundreds	Tens	Ones
8	1	5,	7	9	6

ounce (oz) **[onza (oz)]**

A customary unit used to measure weight

A slice of bread weighs about 1 ounce.

parallel lines **[líneas paralelas]**

Lines that never intersect

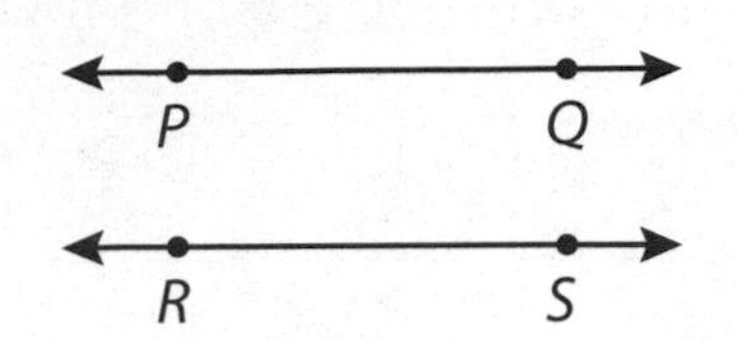

Label: $\overleftrightarrow{PQ} \parallel \overleftrightarrow{RS}$

parallelogram **[paralelogramo]**

A quadrilateral that has two pairs of parallel sides

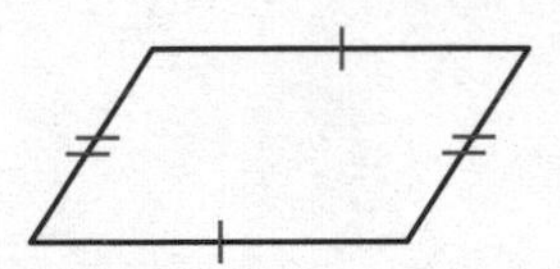

partial products **[productos parciales]**

The products found by breaking apart a factor into ones, tens, hundreds, and so on, and multiplying each of these by the other factor

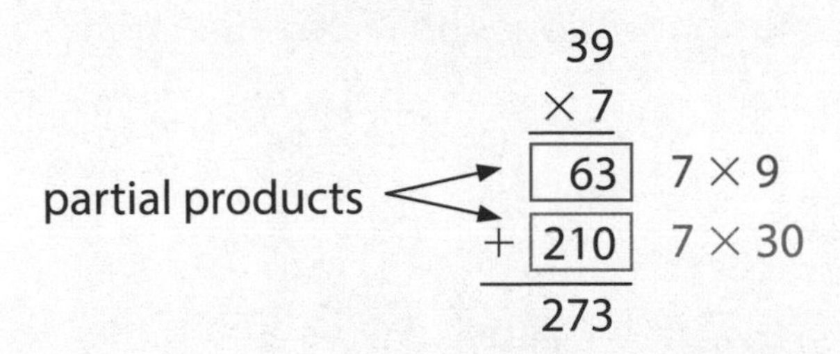

partial quotients **[cocientes parciales]**

A division strategy in which quotients are found in parts until the remainder is less than the divisor

partial quotients

$6\overline{)84}$

$-60 = 6 \times 10$ 10

24

$-24 = 6 \times 4$ $+4$

0   14

perimeter **[perímetro]**

The distance around a figure

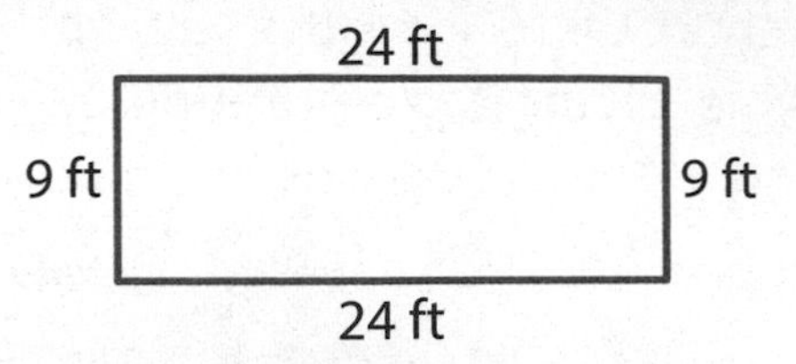

The perimeter is 66 feet.

period **[período]**

Each group of three digits separated by commas in a multi-digit number

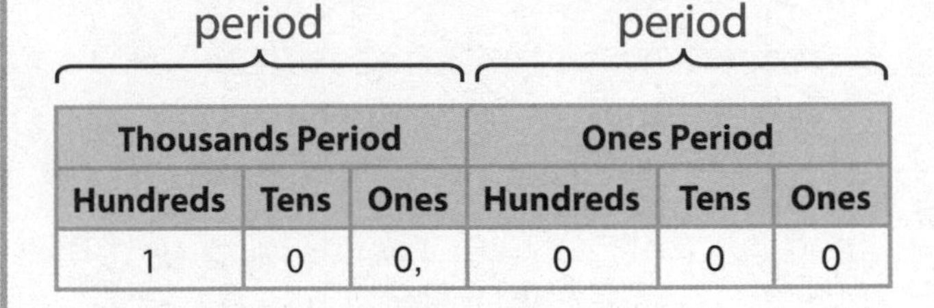

Thousands Period			Ones Period		
Hundreds	Tens	Ones	Hundreds	Tens	Ones
1	0	0,	0	0	0

perpendicular lines **[líneas perpendiculares]**

Lines that intersect to form four right angles

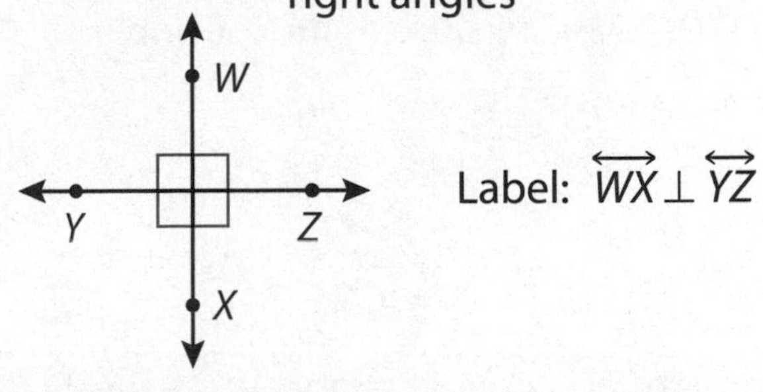

Label: $\overleftrightarrow{WX} \perp \overleftrightarrow{YZ}$

pint (pt) **[pinta (pt)]**

A customary unit used to measure capacity There are 2 cups in 1 pint.

The capacity of the carton is 1 pint.

place value chart
[gráfico de valor posicional]

A chart that shows the value of each digit in a number

Thousands Period			Ones Period		
Hundreds	Tens	Ones	Hundreds	Tens	Ones
2	8	5,	7	4	3

point **[punto]**

An exact location in space

A
•

Label: point A

pound (lb) **[libra (lb)]**

A customary unit used to measure weight
There are 16 ounces in 1 pound.

A loaf of bread weighs about 1 pound.

prime number **[número primo]**

A number greater than 1 with exactly two factors, 1 and itself

11

The factors of 11 are 1 and 11.

protractor **[transportador]**

A tool for measuring and drawing angles

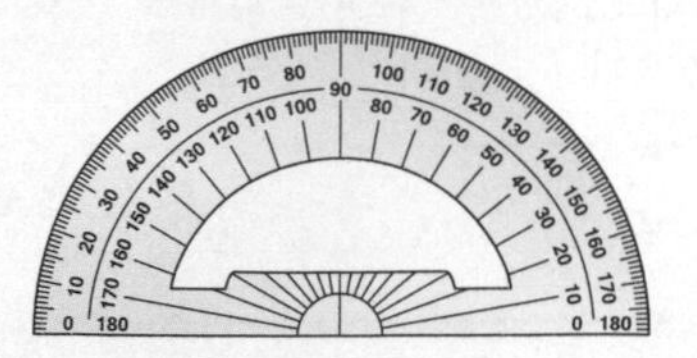

quart (qt) **[cuarto de galón (qt)]**

A customary unit used to measure capacity
There are 2 pints in 1 quart.

The capacity of the carton is 1 quart.

ray **[semirrecta]**

A part of a line that has one endpoint and goes on without end in one direction

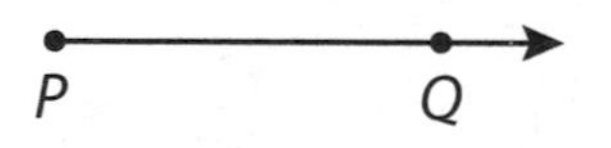

Label: $\overrightarrow{PQ}$

rectangle **[rectángulo]**

A parallelogram that has four right angles

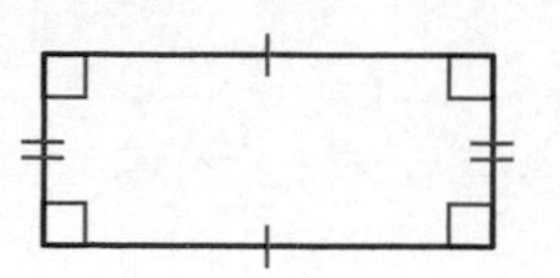

remainder **[resto]**

The amount left over when a number cannot be divided evenly

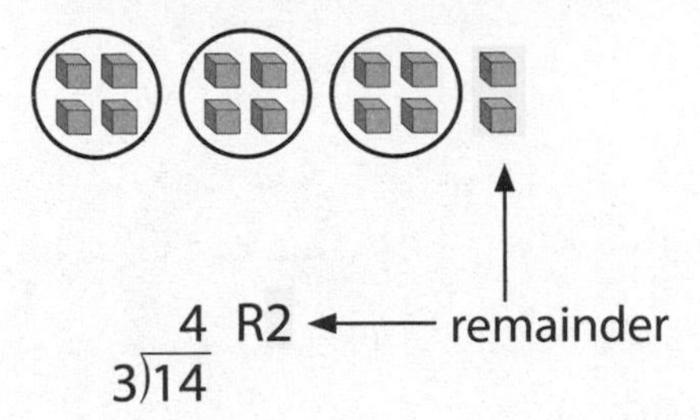

rhombus **[rombo]**

A parallelogram that has four sides with the same length

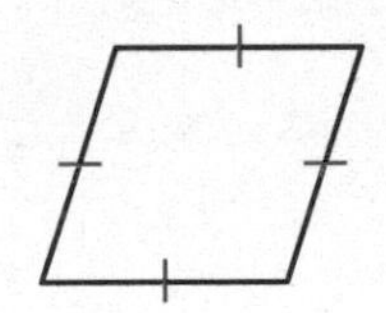

right angle **[ángulo recto]**

An L-shaped angle

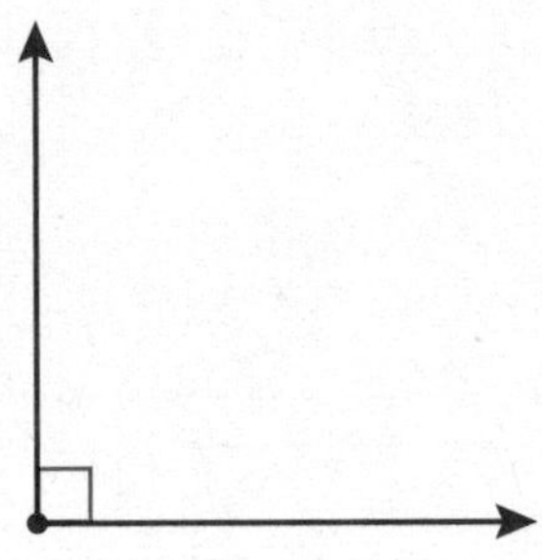

right triangle **[triángulo rectángulo]**

A triangle that has one right angle

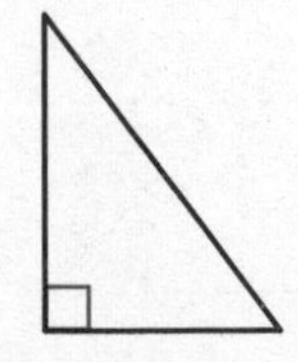

rule **[regla]**

Tells how numbers or shapes in a pattern are related

Rule: Add 3.
3, 6, 9, 12, 15, 18, 21, 24, . . .

Rule: triangle, hexagon, square, rhombus

S

scalene triangle **[triángulo escaleno]**

A triangle that has no sides with the same length

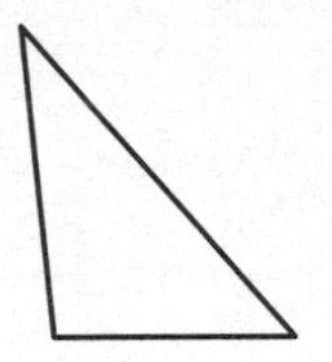

second (sec) **[segundo (seg)]**

A unit of time

1 second

There are 60 seconds in 1 minute.

square **[cuadrado]**

A parallelogram that has four right angles and four sides with the same length

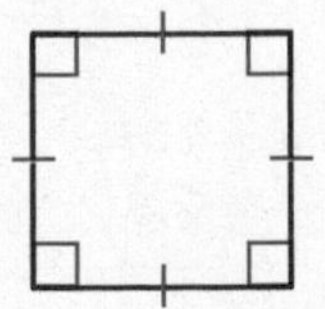

Glossary

straight angle **[ángulo llano]**

An angle that forms a straight line

supplementary angles
[ángulos suplementarios]

Two angles whose measures have a sum of 180°

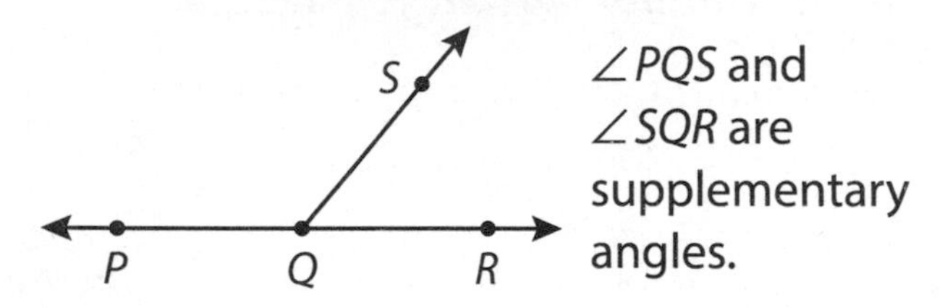

$\angle PQS$ and $\angle SQR$ are supplementary angles.

tenth **[décimo]**

1 of 10 equal parts of a whole

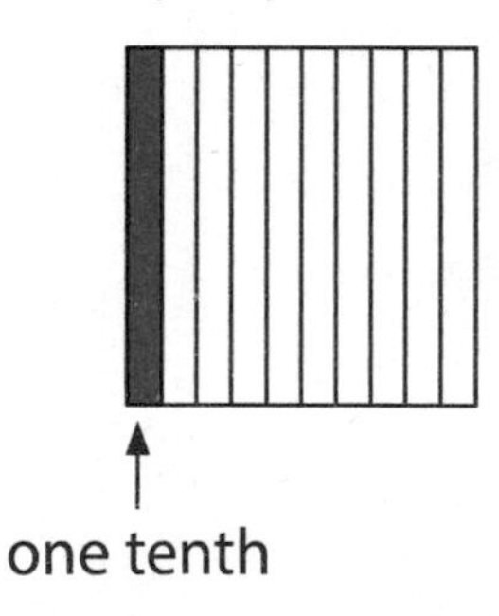

tenths place
[posición de los décimos]

The first place to the right of the decimal point

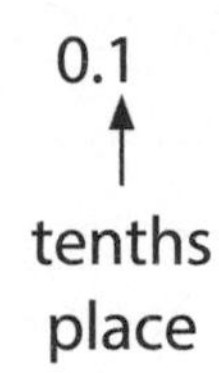

thousands period
[período de las milésimas]

The period after the ones period in a number

Thousands Period			Ones Period		
Hundreds	Tens	Ones	Hundreds	Tens	Ones
8	1	5,	7	9	6

ton (T) **[tonelada (T)]**

A customary unit used to measure weight
There are 2,000 pounds in 1 ton.

A small compact car weighs about 1 ton.

trapezoid **[trapecio]**

A quadrilateral that has exactly one pair of parallel sides

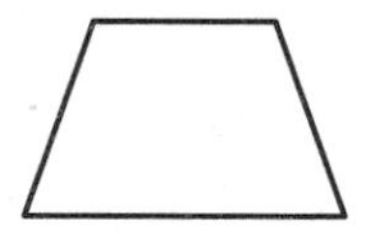

unit fraction **[fracción unitaria]**

Represents one equal part of a whole

Examples:

$\frac{1}{2}$ $\frac{1}{5}$

vertex [vértice]

The endpoint at which two rays or line segments of an angle meet

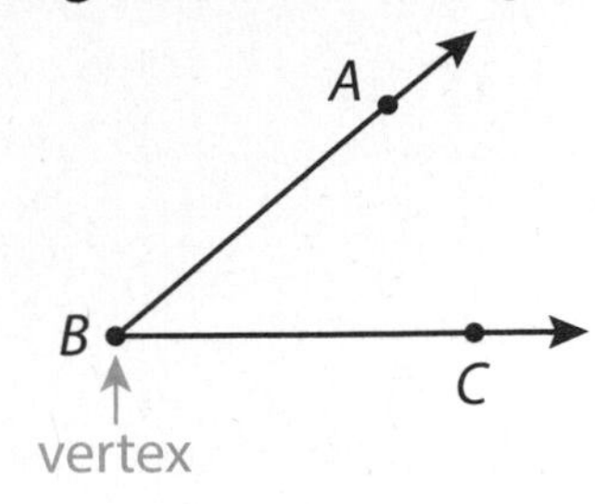

Index

B

C

E

G

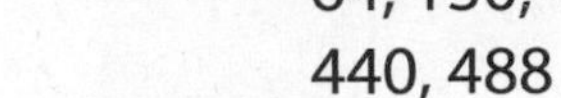

H

M

N

Q

R

S

T

Index

Index

Reference Sheet

Symbols

$\times$	multiply	$^\circ$	degree(s)	$\overleftrightarrow{AB}$	line *AB*
$\div$	divide	$\perp$	is perpendicular to	$\overrightarrow{AB}$	ray *AB*
$=$	equals	$\parallel$	is parallel to	$\overline{AB}$	line segment *AB*
$>$	greater than	$\overset{A}{\bullet}$	point *A*	$\angle ABC$	angle *ABC*
$<$	less than				

Money

¢ cent or cents

$ dollar or dollars

1 penny = 1¢ or $0.01

1 nickel = 5¢ or $0.05

1 dime = 10¢ or $0.10

1 quarter = 25¢ or $0.25

1 dollar ($) = 100¢ or $1.00

Length

Metric

1 centimeter (cm) = 10 millimeters (mm)

1 meter (m) = 100 centimeters

1 kilometer (km) = 1,000 meters

Customary

1 foot (ft) = 12 inches (in.)

1 yard (yd) = 3 feet

1 mile (mi) = 1,760 yards

Mass

1 kilogram (kg) = 1,000 grams (g)

Weight

1 pound (lb) = 16 ounces (oz)

1 ton (T) = 2,000 pounds

Capacity

Metric

1 liter (L) = 1,000 milliliters (mL)

Customary

1 pint (pt) = 2 cups (c)

1 quart (qt) = 2 pints

1 gallon (gal) = 4 quarts

Time

1 second

1 minute (min) = 60 seconds (sec)

1 hour (h) = 60 minutes

1 day (d) = 24 hours

1 week (wk) = 7 days

1 year (yr) = 12 months (mo)

1 year = 52 weeks

Area and Perimeter

Area of a rectangle

$A = \ell \times w$

Perimeter of a rectangle

$P = (2 \times \ell) + (2 \times w)$

length (ℓ)

width (w)

Perimeter of a polygon

P = sum of lengths of sides

Angles

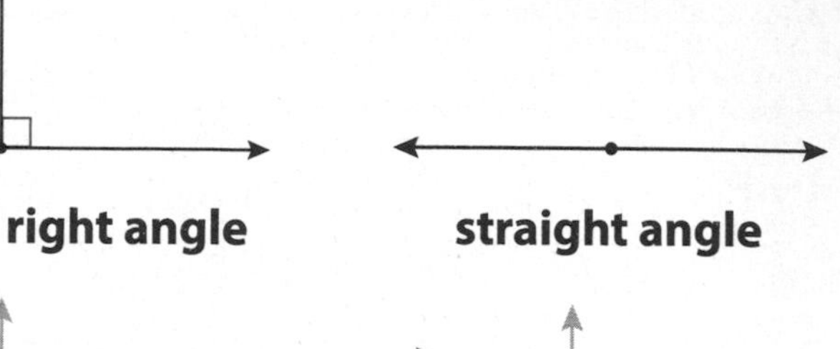

right angle

straight angle

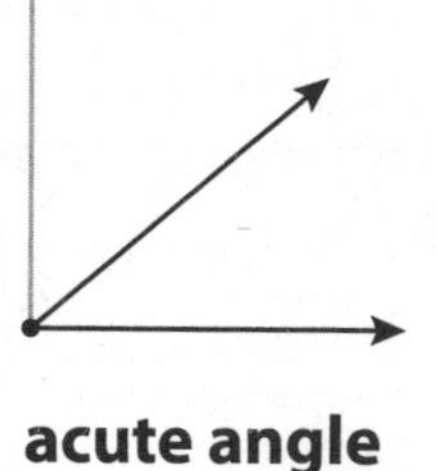

acute angle

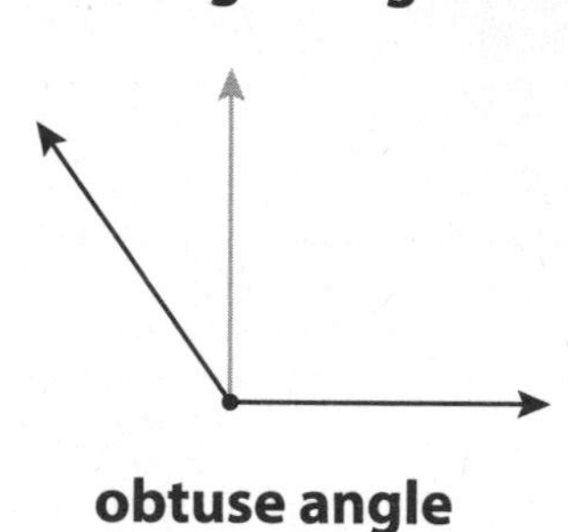

obtuse angle

Triangles

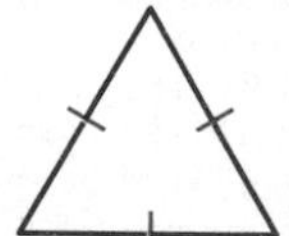

equilateral triangle

isosceles triangle

scalene triangle

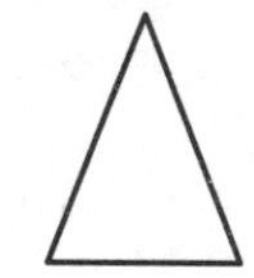

acute triangle

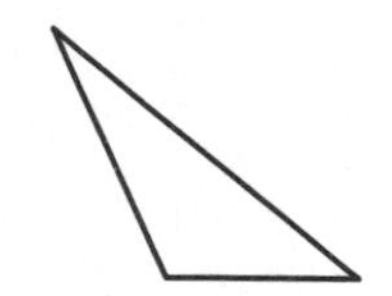

obtuse triangle

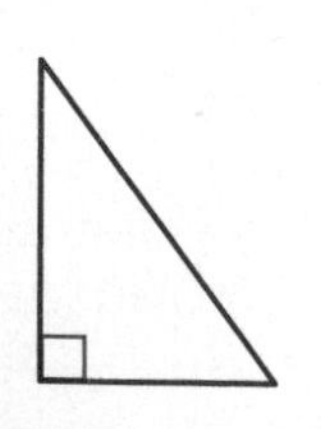

right triangle

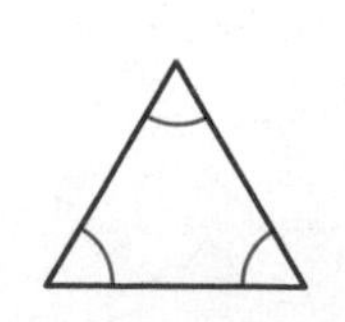

equiangular triangle

Quadrilaterals

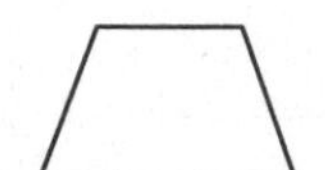

rapezoid

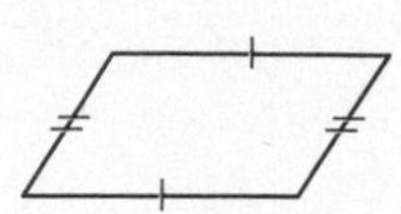

parallelogram

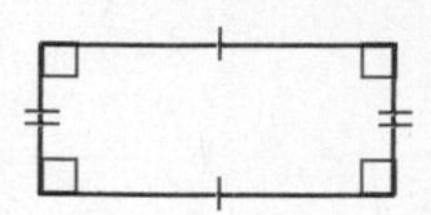

rectangle

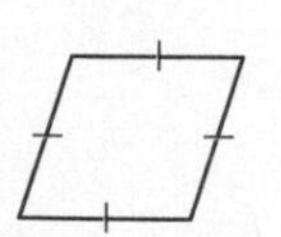

rhombus

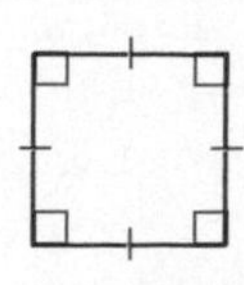

square

Credits

Front matter

i Brazhnykov Andriy /Shutterstock.com; **vii** Steve Debenport/E+/Getty Images

Chapter 1

1 fbxx/iStock/Getty Images Plus; **6** SSSCCC/Shutterstock.com; **8** DanielPrudek/iStock/Getty Images Plus; **11** 3DMI/Shutterstock.com; **12** Lagutkin Alexey/Shutterstock.com; **17** adventtr/iStock/Getty Images Plus; Good Life Studio/E+/Getty Images Plus; **18** Ron_Thomas/E+/Getty Images; **20** ABIDAL/iStock/Getty Images Plus; **24** Rawpixel/iStock/Getty Images Plus; **26** Maren Winter/iStock/Getty Images Plus

Chapter 2

31 SeanPavonePhoto/iStock/Getty Images Plus; **36** *top* grimgram/iStock/Getty Images Plus; Triduza Studio/Shutterstock.com; *bottom* shihina/iStock/Getty Images Plus; **38** ciud/iStock/Getty Images Plus; **41** ET-ARTWORKS/iStock/Getty Images Plus, irin717/iStock/Getty Images Plus; **42** rmbarricarte/iStock/Getty Images Plus; **44** *left* Ann W. Kosche/Shutterstock.com; *right* romakoma/Shutterstock.com; **47** efks/iStock /Getty Images Plus; **48** *left* Rawpixel/iStock/Getty Images Plus; *right* Vladimiroquai/iStock/Getty Images Plus; **50** sndr/E+/Getty Images; **54** *top* vchal/iStock/Getty Images Plus; *Exercise 14* vchal/iStock/Getty Images Plus; *bottom* vchal/iStock/Getty Images Plus; **56** SteffenHuebner/iStock/Getty Images Plus; **59** *top* Brberrys/Shutterstock.com; *bottom* anankkml/iStock/Getty Images Plus; **62** Stubblefield Photography/Shutterstock.com; Todor Rusinov/Shutterstock.com

Chapter 3

67 franckreporter/iStock/Getty Images Plus; **71** Kikkerdirk/iStock/Getty Images Plus; **72** *right* Avesun/iStock/Getty Images Plus; *left* GaryAlvis/E+/Getty Images; **74** *left* catinsyrup/iStock/Getty Images Plus; *right* mphillips007/iStock/Getty Images Plus; **77** malerapaso/E+/Getty Images; **78** *top* k_samurkas/iStock/Getty Images Plus; *bottom* Tempusfugit/iStock/Getty Images Plus; **81** Andy445/E+/Getty Images; Matt Benoit/Shutterstock.com; **83** scanrail/iStock/Getty Images Plus; **84** pictafolio/E+/Getty Images; **90** *right* luminis/iStock/Getty Images Plus; *left* neuson11/iStock/Getty Images Plus; **92** *Exercise 10* AdamParent/iStock/Getty Images Plus; *Exercise 11* cynoclub/iStock/Getty Images Plus; **96** *top* Goddard_Photography/iStock/Getty Images Plus; *bottom* thawats/iStock/Getty Images Plus; **98** *top* CathyKeifer/iStock/Getty Images Plus; *bottom* Steve Collender/Shutterstock.com; **102** *top* lorcel/iStock/Getty Images Plus; *bottom* W6/iStock/Getty Images Plus; **104** *Exercise 12* vencavolrab/iStock/Getty Images Plus; *Exercise 13* adogslifephoto/iStock/Getty Images Plus; **108** *top* Ververidis Vasilis/Shutterstock.com; *bottom left* YinYang/iStock/Getty Images Plus; *bottom right* filrom/iStock/Getty Images Plus, Vereshchagin Dmitry/Shutterstock.com; **110** *Exercise 15* ID1974/Shutterstock.com; *Exercise 16* Nerthuz /Shutterstock.com; **113** paylessimages/iStock/Getty Images Plus; **114** *top right* leksele/iStock/Getty Images Plus; David Osborn/Shutterstock.com; *Exercise 18* itographer/E+/Getty Images; IPGGutenbergUKLtd/iStock/Getty Images Plus; **120** *top* Carso80/iStock Editorial/Getty Images Plus; *center* D3Damon/iStock/Getty Images Plus; *bottom* Iakov Filimonov/Shutterstock.com; 4kodiak/iStock Unreleased/Getty Images Plus; **122** *left* 4kodiak/iStock Unreleased/Getty Images Plus; *right* Aerotoons/DigitalVision Vectors/Getty Images; **123** Mirko Rosenau/Shutterstock.com; **125** *top left* GlobalP/iStock/Getty Images Plus; *top right* Antagain/iStock/Getty Images Plus; *bottom* roberthyrons/iStock/Getty Images Plus; **126** PeopleImages/E+/Getty Images; **128** kropic1/Shutterstock.com, ©iStockphoto.com/Chris Schmidt, ©iStockphoto.com/Jane norton; **129** *top* Maxiphoto/iStock/Getty Images Plus; *bottom* dibrova/iStock/Getty Images Plus; **130** mocoo/iStock/Getty Images Plus; ONYXprj/iStock/Getty Images Plus; Sylphe_7/iStock/Getty Images Plus; **134** magnez2/iStock Unreleased/Getty Images Plus; **136** popovaphoto/iStock/Getty Images Plus; **139** *top* TimBoers/iStock/Getty Images Plus; *bottom* Nielskliim/Shutterstock.com; **140** Wiese_Harald/iStock/Getty Images Plus

Chapter 4

141 Liufuyu/iStock/Getty Images Plus; **145** JuSun/iStock/Getty Images Plus; **146** VStock/Alamy Stock Photo; **148** Steve Collender /Shutterstock.com; **152** GlobalP/iStock/Getty Images Plus; **157** bjdlzx/iStock/Getty Images Plus; **158** *top* spanteldotru/E+/Getty Images; *bottom* AlbertoRoura/iStock Editorial/Getty Images Plus; **160** mipan/iStock/Getty Images Plus; **164** *top* bortonia/DigitalVision Vectors/Getty Images; *bottom* artisteer/iStock/Getty Images Plus, vectorloop/DigitalVision Vectors/Getty Images; **166** Thor Jorgen Udvang /Shutterstock.com; **169** GlobalP/iStock/Getty Images Plus; **170** *top* martinhosmart/iStock/Getty Images Plus; *Exercise 13* ianmcdonnell/E+/Getty Images; **176** *top* ChrisGorgio/iStock/Getty Images Plus; *Exercise 13 right* Konstantin G/Shutterstock.com; *Exercise 13 left* Triduza/iStock/Getty Images Plus; *bottom left* John_Kasawa/iStock/Getty Images Plus; **178** leezsnow/iStock/Getty Images Plus; **182** *top* Isaac74/iStock Editorial/Getty Images Plus; *bottom* breckeni/E+/Getty Images; **184** Jamesmcq24/E+/Getty Images; **185** Blade_kostas/iStock/Getty Images Plus; **186** tanyasharkeyphotography/iStock/Getty Images Plus; **187** *Exercise 4* creatOR76/Shutterstock.com; *Exercise 6* ZU_09/DigitalVision Vectors/Getty Images; **188** THEPALMER/E+/Getty Images; **189** *top* macrovector/iStock/Getty Images Plus; *bottom* RelaxFoto.de/E+/Getty Images; **190** *Exercise 6* filo/DigitalVision Vectors/Getty Images; *bottom* Aun Photographer/Shutterstock.com; **191** vladru/iStock/Getty Images Plus; **196** *Exercise 45* Believe_In_Me/iStock/Getty Images Plus; *Exercise 46* BrianAJackson/iStock/Getty Images Plus

Chapter 5

197 BanksPhotos/iStock/Getty Images Plus; **201** Grigorenko/iStock/Getty Images Plus; **202** *top* skynesher/E+/Getty Images; *bottom* quavondo/iStock/Getty Images Plus; **204** GlobalP/iStock/Getty Images Plus; **207** mar1koff/iStock/Getty Images Plus; **208** *top* Antonio V. Oquias /Shutterstock.com; *bottom* dageldog/iStock/Getty Images Plus; **210** *top* Twoellis/iStock/Getty Images Plus; *Exercise 10 left* ahirao_photo/iStock/Getty Images Plus; *Exercise 10 right* mtruchon/iStock/Getty Images Plus; **214** *top* Peter Hermus/iStock/Getty Images Plus; *bottom* kroach/iStock/Getty Images Plus; **216** duckycards/E+/Getty Images; **220** *top* Valerie Loiseleux/iStock/Getty Images Plus; *bottom* kali9/Vetta/Getty Images; **222** cynoclub/iStock/Getty Images Plus; **225** FatCamera/E+/Getty Images; **226** *top* FatCamera/E+/Getty Images; *Exercise 18* ONYXprj/iStock/Getty Images Plus; **228** Mathisa_s/iStock/Getty Images Plus; **232** *top* Voren1/iStock/Getty Images Plus; *Exercise 18* Goce Risteski/Hemera/Getty Images; **234** *top* sarent/iStock/Getty Images Plus; *Exercise 13* tashka2000/iStock/Getty Images Plus; **237** ziherMP/iStock/Getty Images Plus; **238** *top* CTRPhotos/iStock Editorial/Getty Images Plus; *bottom* sudok1/iStock/Getty Images Plus; **240** *top* Cgissemann/iStock/Getty Images Plus; *Exercise 15* WildDoc/iStock Unreleased/Getty Images; **243** humonia/iStock/Getty Images Plus; **244** McIninch/iStock/Getty Images Plus; **246** aluxum/iStock/Getty Images Plus; **247** kaanates/iStock/Getty Images Plus; **248** artplay711/iStock/Getty Images Plus; **249** *top* Monkeybusinessimages/iStock/Getty Images Plus; *bottom* Onandter_sean/iStock/Getty Images Plus; **250** AdamRadosavljevic/iStock/Getty Images Plus; **251** *left* AndreaAstes/iStock/Getty Images Plus; *right* OlegAlbinsky/iStock/Getty Images Plus; **252** *top* EdnaM/iStock/Getty Images Plus; *bottom* USO/iStock/Getty Images Plus; **253** Werner Otto/Alamy Stock Photo; **256** vladimir_n/iStock/Getty Images Plus; **258** shark_749/iStock/Getty Images Plus

Chapter 6

259 monkeybusinessimages/iStock/Getty Images Plus; **264** *top right* crossbrain66/E+/Getty Images, karandaev/iStock/Getty Images Plus; *Exercise 15* SeaHorseTwo/iStock/Getty Images Plus; **266** mipan/iStock/Getty Images Plus; **270** ivanmateev/iStock/Getty Images Plus; **272** Diana Taliun/iStock./Getty Images Plus; **276** *top* Balkonsky/Shutterstock.com; *bottom* pagadesign/E+/Getty Images; **282** *top* BruceBlock/iStock/Getty Images Plus; *bottom* brozova/iStock/Getty Images Plus; **284** kutaytanir/E+/Getty Images; **288** *top* nojustice/E+/Getty Images; *bottom* myistock88/iStock/Getty Images Plus; **290** Evgeny555/iStock/Getty Images Plus; **297** *top* leezsnow/iStock/Getty Images Plus; *center* elinedesignservices/iStock/Getty Images Plus; *bottom* stevezmina1/DigitalVision Vectors/Getty Images; **301** labsas/iStock/Getty images Plus; 172969371/E+/Getty Images; ar-chi/iStock/Getty Images Plus

Chapter 7

303 kali9/iStock/Getty Images Plus; **307** stockcam/iStock/Getty Images Plus; **308** *top* Zolotaosen/iStock/Getty Images Plus; *center* repinanatoly/iStock/Getty Images Plus; *bottom* wavebreakmedia/Shutterstock.com; **310** AYImages/E+/Getty Images; **314** *top* sarahdoow/iStock/Getty Images Plus; *bottom* France68/iStock/Getty Images Plus; **316** Elnur Amikishiyev/Hemera/Getty Images Plus; stuartbur/iStock/Getty Images Plus; **320** Nikada/iStock/Getty Images Plus; **322** *left* Eyematrix/iStock/Getty Images Plus; *right* RimDream/iStock/Getty Images Plus; **325** Remus86/iStock/Getty Images Plus; **326** JoKMedia/iStock/Getty Images Plus; **328** dem10/iStock/Getty Images Plus; Dixi_/iStock/Getty Images Plus; **332** *top* suksao999/iStock/Getty Images Plus; *bottom* Krasyuk/iStock/Getty Images Plus; **334** mgkaya/iStock/Getty Images Plus; **337** AlexStar/iStock/Getty Images Plus; valery121283/iStock/Getty Images Plus; **339** pressureUA/iStock /Getty Images Plus, Rawpixel/iStock /Getty Images Plus; **340** OLEG525/iStock/Getty Images Plus; **342** kali9/E+/Getty Images; **343** GCShutter/E+/Getty Images

Chapter 8

345 FatCamera/E+/Getty Images; **350** *top* fstop123/E+/Getty Images; *bottom 15* Cheng Wei/Shutterstock.com; **356** *top* JodiJacobson/iStock/Getty Images Plus; *bottom*; Emevil/iStock/Getty Images Plus; **358** flas100/iStock/Getty Images Plus; **362** *top* Lonely__/iStock/Getty Images Plus; *bottom* etvulc/iStock/Getty Images Plus; **368** *top* pr2is/iStock/Getty Images Plus; *bottom* RonBailey/iStock/Getty Images Plus; **370** rossandgaffney/iStock/Getty Images Plus; **373** chictype/E+/Getty Images; **374** sharply_done/E+/Getty Images; **380** Quality-illustrations/iStock/Getty Images Plus; **386** Alter_photo/iStock/Getty Images Plus; **388** filo/E+/Getty Images; **392** *left* AndreaAstes/iStock/Getty Images Plus; *right* Lawrence Manning/Corbis/Getty Images; **395** *top* mazzzur/iStock/Getty Images Plus; *bottom* Taalvi/iStock/Getty Images Plus; **396** solarseven/iStock/Getty Images Plus; **397** *top* ana belen prego alvarez/iStock/Getty Images Plus; *bottom* MR1805/iStock/Getty Images Plus; **398** *top* onurdongel/iStock/Getty Images Plus; *bottom* bluestocking/E+/Getty Images; **400** PIKSEL/iStock/Getty Images Plus

Chapter 9

407 georgeclerk/E+/Getty Images; **412** *top* timsa/iStock/Getty Images Plus; *bottom* EHStock/iStock/Getty Images Plus; **414** *top* anna1311/iStock/Getty Images Plus; *bottom* AbbieImages/iStock/Getty Images Plus; **418** *top* cris180/iStock/Getty Images Plus; *bottom* StockPhotosArt/iStock/Getty Images Plus; **420** Scientifics Direct/www.ScientificsOnline.com; **424** *top* Jupiterimages/PHOTOS.com>>/Getty Images Plus; *bottom* mbbirdy/iStock Unreleased / Getty Images Plus; **430** *top* atosan/iStock/Getty Images Plus; *bottom* wickedpix/iStock/Getty Images Plus; **432** Scrambled/iStock/Getty Images Plus; **433** filo/DigitalVision Vectors/Getty Images; **434** eriksvoboda/iStock/Getty Images Plus; **435** fcafotodigital/E+/Getty Images; **436** Elenathewise/iStock/Getty Images Plus; **437** margouillatphotos/iStock/Getty Images Plus; **438** FernandoAH/E+/Getty Images; **439** mm88/iStock/Getty Images Plus

Chapter 10

443 StephanieFrey/iStock/Getty Images Plus; **444** TokenPhoto/E+/Getty Images; **447** berean/iStock/Getty Images Plus; **448** *top* Juanmonino/iStock/Getty Images Plus; *bottom* JamesBrey/E+/Getty Images; **450** *Exercise 14* master1305/iStock/Getty Images Plus; *Exercise 16* nicolamargaret/E+/Getty Images; **453** *right* sumos/iStock/Getty Images Plus; *left* OSTILL/iStock/Getty Images Plus; **454** *top* DmitriyKazitsyn/iStock/Getty Images Plus; *bottom* skegbydave/iStock/Getty Images Plus; **456** *top left* mrPliskin/E+/Getty Images; *top right* kyoshino/E+/Getty Images; *Exercise 21* Nearbirds/iStock/Getty Images Plus, dosrayitas/DigitalVision Vectors/Getty Image; **460** *top* Coprid/iStock/Getty Images Plus; *bottom* banprik/iStock/Getty Images Plus; **462** *left* Donald Miralle/DigitalVision/Getty Images; *right* FatCamera/iStock/Getty Images Plus; **466** nidwlw/iStock/Getty Images Plus **472** *top* ET1972/iStock/Getty Images Plus; *bottom* Imgorthand/E+/Getty Images; **474** *top* GlobalP/iStock/Getty Images Plus; *bottom* Chris Mattison / Alamy Stock Photo; **478** *right* Oakozhan/iStock/Getty Images Plus; *left* FuzzMartin/iStock/Getty Images Plus; **480** esanbanhao/iStock/Getty Images Plus; **483** popovaphoto/iStock/Getty Images Plus; **484** ChrisGorgio/iStock/Getty Images Plus; kbeis/DigitalVision Vectors/Getty Images; **485** ryasick/iStock/Getty Images Plus; **486** *Exercise 5* ryasick/iStock/Getty Images Plus; *Exercise 7 left* Africa Studio/Shutterstock.com; *Exercise 7 right* AlexLMX/iStock/Getty Images Plus; **487** *right* SednevaAnna/iStock/Getty Images Plus; *left* eduardrobert/iStock/Getty Images Plus; **89** DebbiSmirnoff/iStock/Getty Images Plus; **492** *left* phantq/iStock/Getty ges Plus; *right* esseffe/iStock/Getty Images Plus

Chapter 11

493 kali9/E+/Getty Images; **494** *left* PLAINVIEW/E+/Getty Images; *center* cmannphoto/iStock/Getty Images Plus; *right* wwing/iStock/Getty Images Plus; **494b** *From left to right* Ryan McVay/Photodisc; sarahdoow/iStock/Getty Images Plus; SerrNovik/iStock /Getty Images Plus; antpkr/iStock/Getty Images Plus; Canadapanda/iStock/Getty Images Plus; whitewish/iStock/Getty Images Plus; Hemera Technologies/PhotoObjects.net/Getty Images Plus; youngID/DigitalVision Vectors/Getty Images; **498** *top* FatCamera/iStock/Getty Images Plus; *Exercise 18 left* Rvo233/iStock/Getty Images Plus; *Exercise 18 right* Onfokus/E+/Getty Images; *bottom* TonyTaylorStock/iStock Editorial/Getty Images Plus; **500** DougBennett/E+/Getty Images; **502** ikitano/iStock/Getty Images Plus; **503** Marakit_Atinat/iStock/Getty Images Plus; **504** PicturePartners/iStock/Getty Images Plus; **510** *left* amattel/iStock/Getty Images Plus; *right* cellistka/iStock/Getty Images Plus; **514** Blade_kostas/iStock/Getty Images Plus; **516** *top* Cloudtail_the_Snow_Leopard/iStock/Getty Images Plus; *bottom* DarthArt/iStock/Getty Images Plus; **518** GlobalP/iStock/Getty Images Plus; **522** Elenathewise/iStock/Getty Images Plus; **524** Photograpе/iStock/Getty Images Plus; **525** sam74100/iStock/Getty Images Plus; **527** Farinosa/iStock/Getty Images Plus; **529** Hreni/iStock/Getty Images Plus; **530** Turnervisual/iStock/Getty Images Plus; **534** JackF/iStock/Getty Images Plus; **539** Madmaxer/iStock/Getty Images Plus; **541** kali9/E+/Getty Images; **542** JonathanCohen/E+/Getty Images; **545** skaljac/Shutterstock.com; **546** jdwfoto/iStock Unreleased/Getty Images Plus; **548** Erdosain/iStock/Getty Images Plus; **549** Rich Vintage/Vetta/Getty Images; **551** mbongorus/iStock/Getty Images Plus; **552** Mastervision/Bi-Silque; **553** scubaluna/iStock/Getty Images Plus; **554** samards/iStock/Getty Images Plus; **558** Colin_Davis/iStock/Getty Images Plus; **559** Krasyuk/iStock/Getty Images Plus; **560** Pixsooz/iStock/Getty Images Plus

Chapter 12

561 dsabo/iStock/Getty Images Plus; **566** *left* finevector/iStock/Getty Images Plus; *right* ensieh1/iStock/Getty Images Plus; **568** iconogenic/iStock/Getty Images Plus; **572** *top* Eyematrix/iStock/Getty Images Plus; *bottom* GrafVishenka/iStock/Getty Images Plus; **574** denisik11/iStock/Getty Images Plus; **578** *Exercise 11* nonnie192/iStock/Getty Images Plus; *Exercise 12* Krimzoya/iStock/Getty Images Plus; **581** ismagilov/iStock/Getty Images Plus; **583** *Exercise 5* marekuliasz/iStock/Getty Images Plus; *bottom* VvoeVale/iStock/Getty Images Plus; **584** kpalimski/iStock/Getty Images Plus; **587** stocknshares/E+/Getty Images, paci77/E+/Getty Images; **590** elinedesignservices/iStock/Getty Images Plus

Chapter 13

591 JaysonPhotography/iStock/Getty Images Plus; **610** elinedesignservices/iStock/Getty Images Plus; **614** *top* Vitalliy/iStock/Getty Images Plus; *center* MicrovOne/iStock/Getty Images Plus; *bottom* szefei/iStock/Getty Images Plus; **616** Orla/iStock/Getty Images Plus; **620** *top* AnatolyM/iStock/Getty Images Plus; *center* marimo_3d/iStock/Getty Images Plus; *bottom* olgna/iStock/Getty Images Plus; **622** jonathansloane/E+/Getty Images, OLEKSANDR PEREPELYTSIA/iStock/Getty Images Plus; **634** iZonda/iStock/Getty Images Plus; **638** pialhovik/iStock/Getty Images Plus

Chapter 14

647 suebee65/iStock/Getty Images Plus; **655** adekvat/iStock/Getty Images Plus; VectorPocket/iStock/Getty Images Plus; **664** marekuliasz/iStock/Getty Images Plus; **665** *Exercise 4* Carol_Anne/iStock/Getty Images Plus; *Exercise 6* StockImages_AT/iStock/Getty Images Plus; **669** *Exercise 9* lucielang/iStock/Getty Images Plus; *Exercise 10* blue64/E+/Getty Images; *Exercise 11* Marje4/E+/Getty Images; **670** photovideostock/E+/Getty Images; **671** *Exercise 7* SlidePix/iStock/Getty Images Plus; *Exercise 8* ibooo7/iStock/Getty Images Plus; **675** *Exercise 5* Casey Botticello/iStock/Getty Images Plus; *Exercise 6* stevezmina1/DigitalVision Vectors/Getty Images; **677** antpkr/iStock/Getty Images Plus; **681** egal/iStock/Getty Images Plus; **687** Vladimir/iStock/Getty Images Plus; **688** *top* snowpeace19/iStock/Getty Images Plus; *bottom* Vladimir/iStock/Getty Images Plus

Cartoon Illustrations: MoreFrames Animation
Design Elements: oksanika/Shutterstock.com; icolourful/Shutterstock.com; Valdis Torms